WERTHEIM FELLOWSHIP
PUBLICATIONS
HARVARD ECONOMIC STUDIES
VOLUME XCIV

THE
NEW ENGLAND
FISHING INDUSTRY

A STUDY IN PRICE AND WAGE SETTING

Donald J. White

HARVARD UNIVERSITY PRESS
Cambridge, Massachusetts
1954

LIBRARY OF CONGRESS CATALOG CARD NUMBER 54–7065

PRINTED IN THE UNITED STATES OF AMERICA

WERTHEIM FELLOWSHIP PUBLICATIONS

In 1923 the family of the late Jacob Wertheim established the Jacob Wertheim Research Fellowship for ". . . the support of original research in the field of industrial coöperation . . ." The Fellowship was intended to enable men and women ". . . who already have expert knowledge of this subject, to pursue research that may be of general benefit in solving the problems in this field . . ." Fellowships are awarded annually by the President and Fellows of Harvard College on the recommendation of the Wertheim Committee.

The Committee undertakes to provide general supervision to the program of research of the Wertheim Fellow. When that research yields findings and results which are significant and of general interest, the Committee is authorized by the terms of the grant to Harvard University to recommend publication. The Jacob Wertheim Research Fellow alone has responsibility for the facts, analysis, and opinions expressed in this volume.

John D. Black, *Chairman*
Sumner H. Slichter
B. M. Selekman
Samuel A. Stouffer
John T. Dunlop, *Secretary*

WERTHEIM FELLOWSHIP PUBLICATIONS

Donald J. White, *The New England Fishing Industry: A Study in Price and Wage Setting*, 1954

Theodore V. Purcell, S.J., *The Worker Speaks His Mind on Company and Union*, 1953

Lloyd H. Fisher, *The Harvest Labor Market in California*, 1953

Walter Galenson, *The Danish System of Labor Relations: A Study in Industrial Peace*, 1952

John T. Dunlop and Arthur D. Hill, *The Wage Adjustment Board: Wartime Stabilization in the Building and Construction Industry*, 1950

Ralph Altman, *Availability for Work: A Study in Unemployment Compensation*, 1950

Dorothea de Schweinitz, *Labor and Management in a Common Enterprise*, 1949

Walter Galenson, *Labor in Norway*, 1949

Leo C. Brown, S.J., *Union Policies in the Leather Industry*, 1947

Paul H. Norgren, *The Swedish Collective Bargaining System*, 1941

Johnson O'Connor, *Psychometrics*, 1934

William Haber, *Industrial Relations in the Building Industry*, 1930

Wertheim Lectures on Industrial Relations, 1929

J. D. Houser, *What the Employer Thinks*, 1927

FOREWORD

This volume provides a balanced and a comprehensive treatment of the problems of the New England fishing industry. It will interest alike policy makers in business, unions, state and federal government agencies, and Professor White's academic colleagues. As the subtitle indicates, this is a study in the interrelations of wages and prices and factor and product markets.

The compensation of fishermen depends directly upon a formula for splitting the revenue derived from the sale of raw fish (after deducting certain expenses) between the crew of the vessel and the owners. This method of wage payment inevitably tends to involve the fishermen and their union in ways of influencing the sales of raw fish: the pricing arrangements used to set the price of the fish, the size of the catch landed, and the methods of determining the quality of the fish. The division of sales between the crew and the owners seems at times to constitute a "built-in war."

The Theoretical Supplement will be of interest to economists as it constitutes a contribution in the application of theory to interrelated labor and product markets.

The chapters which analyze collective bargaining in this industry illuminate the process by which parties adapt their organization and their policies to the problems of their particular environment. While Professor White recognizes that the union faces the task of reconciling internal differences, the study shows convincingly that the policies developed by the union are designed to secure the greatest possible income for its members. The author accepts this objective for the union but suggests that policies designed to increase the demand for fish would yield even higher wage incomes.

The study has the great merit that it goes beyond theoretical analysis and a report on the operation of collective bargaining to consider specific proposals for the New England fishing industry. These suggestions are based on a real understanding of the industry and its problems and deserve most careful consideration by all interested parties. It is clear that the future of this historic industry in New England depends upon a reappraisal of its techniques, methods, and organization. Professor White has done a service in placing these possibilities in perspective.

JOHN T. DUNLOP

CONTENTS

EXHIBITS

THEORETICAL SUPPLEMENT

AUTHOR'S PREFACE

No natural resource is richer in history than the New England fisheries. Indeed, they constitute the basis of the oldest commercial enterprise in our entire country. The founders of Plymouth Colony, for example, came to America to serve God — and to fish. In the seventeenth and eighteenth centuries fishing was the cornerstone of colonial power. Export trade in salt fish provided the foundation for New England's mounting maritime commerce. The fishing industry and the trade it helped make possible were well-springs of capital for the surge of manufacturing growth which early made New England America's workshop. It is no accident that for generations a "sacred cod" has been displayed prominently in the Massachusetts House of Representatives.

For many decades now, however, the industry no longer has been the most important in the region. Today, in all its major divisions, it provides income directly for probably not more than fifty thousand of New England's four and one quarter million work force. Nevertheless, the industry remains a natural-resource-oriented endeavor in an area notably short of resource-based enterprise.

This study is concerned with but one of the industry's major divisions, the fresh and frozen finny fish sector concentrated in Boston, Gloucester, and New Bedford, Massachusetts; and in Portland and Rockland, Maine. In focusing the spotlight on operations in these ports, we are purposely excluding from coverage such other prominent branches of the trade as sardine canning in Maine, shellfishing and lobstering in all the coastal states (but particularly in Maine and Massachusetts), and brokerage in all kinds of fish specialties, which is centered in Boston. Each of these, indeed, is a well knit "industry" suited for separate study.

Our analysis focuses upon the Atlantic Fishermen's Union, relations between that organization and vessel owners and fish buyers, and the policies these groups have developed and carried forward. The discussion is developed in the light of the basic economic characteristics and problems of the industry. Most noteworthy is the fact that the fishermen work under a system of wage payment called the *lay*. With this arrangement, the fishermen share in the value of the catch and the expenses of each fishing trip. Thus the lay is a "pure profit-sharing" device in the sense that the fishermen's fortunes have turned completely on the success of each voyage. Only in Boston and to a very minor extent in New Bedford have the fishermen had any income guarantees. The lay has antedated and gone much farther than those profit-sharing arrangements, growing in number in American industry, where workers receive regular wages and share additionally in the profits of the enterprise when they are earned.

Through their union, the fishermen have bargained with the vessel owners to determine the respective share which each shall have in the value of the catch and to determine which of the operating expenses of each trip the fishermen shall meet from their share and which of those expenses shall be borne, in part at least, by the owner. The fishermen and their union come into contact with the fish buyers in the major ports because the prices offered by those buyers in primary markets determine the value of any given catch upon which the crew's earnings depend. We have attempted to examine critically the policies developed by each major interest group, and particularly those installed by the union to insure satisfactory incomes. The emphasis is upon explaining the origin and rationale of these policies in terms of the human and economic forces at work. Why have the fishermen at times deliberately limited production? Why have the buyers resisted pressures of the Union to change primary market price-making arrangements? Have the stands taken by the respective groups been wise, when looked at from the standpoint of the goals each wishes to achieve? These are some of the questions we have attempted to answer. In dealing with these matters, we have woven into the study an examination of the key problems of fish scarcity, fish marketing, and foreign competition which harass fishermen, vessel owners, and fish buyers alike. Finally, we have offered some suggestions which might help the parties minimize their internal differences and overcome their common problems, and we have reviewed the prospects for progress.

It is hoped that the analysis and the recommendations founded upon it will prove of value to each of the major economic groups.

Advice is much easier to hand out than to take. Moreover, critical analysis particularly is often unpalatable. Nonetheless, the author will be deeply gratified if the study succeeds in some measure in promoting moves and arrangements which strengthen the industry.

So many thanks are due so many people for the part they have had in this endeavor that we hesitate to name them for fear of missing someone. However, in the pages that follow, an attempt is made to include them all. Without their aid, the work could not have been completed. Particularly, we wish to thank Captain Patrick McHugh of the Atlantic Fishermen's Union who made available the files of that organization and gave unlimited time in interviews; Mr. Thomas D. Rice of the Massachusetts Fisheries Association, who allowed inspection of many of his records and volunteered much information; Mr. Edmund O'Neil, former secretary of the New Bedford Seafood Producers' Association, who provided much information on that port and made it possible to interview many of the port's leaders; and Mr. L. J. Hart of the Gloucester Fisheries Association who helped greatly with materials and arranging interviews. Much of the historical material was made available by the late John C. Wheeler, formerly of the Sales Department, General Seafoods Division, General Foods Corporation in Boston. Before his death, Mr. Wheeler carefully reviewed much of the manuscript for technical correctness. Not enough can be said for the coöperation given by the officials of the U.S. Fish and Wildlife Service and their counterparts at the state level in Massachusetts and Maine. Professor John T. Dunlop of Harvard University, who suggested the study originally as a project for a doctoral dissertation, made invaluable comments and suggestions and constantly encouraged the effort. Professor Sumner H. Slichter, too, lent much encouragement, proferred many ideas, and allowed the opportunity to present parts of the work for discussion at his Collective Bargaining Seminar at Harvard University. Dr. Alfred C. Neal of the Federal Reserve Bank of Boston made possible the publication of an article based on this research in the bank's *Monthly Review* of March, 1950. Both Dr. Neal and Dr. Arthur A. Bright Jr. of the bank made many excellent suggestions. Parts of the author's research were used by Dr. Bright and his staff in the preparation of the fisheries section of the Committee of New England's report on the region. Some of the ideas advanced here have appeared also in the report on New England prepared in 1950 for the President's Council of Economic Advisers, a project on which the author was privileged to coöperate.

In the pages that follow are listed the names and affiliations of others now or formerly associated with the industry who lent a helping hand. In Boston, E. L. Dunn and Frank Smith, New England Fish Exchange; Joseph Delay, Boston Fish Market Corporation; Joseph O'Brien, National Fish Division, A & P Tea Co.; Thomas A. Fulham, Fulham Bros.; H. S. Trilling, Bonnie Fisheries; E. H. Cooley, formerly of Atlantic Coast Fisheries; Thomas Cummings, F. J. O'Hara and Sons; R. M. Wagner, John Kaylor, W. F. Hampton, General Foods; M. J. Dillon and Joseph Cahill, formerly of General Foods; James Carlson, Baker, Boies, and Watson; Henry Wise, Counsel, Daniel MacMillan, and James Quinn, Atlantic Fishermen's Union; John Donegan, Seafood Workers' Union. In Gloucester, F. McG. Bundy, H. H. Bundy, Jr., W. W. Sellew, Gorton Pew Fisheries Co.; T. E. Burns, formerly of Gorton Pew Fisheries Co.; John Del Torchio, Cape Ann Fisheries; J. J. Burke, Jr., M. and B. Trawling Co.; M. I. Bernstein, Gloucester Fishing Vessel Owners Association; John Cahill, Attorney, Gloucester Fisheries Association; Manuel "Jeff" Domingos, United Fisheries; A. F. Hayes, George Hodgson, and John Muise, Atlantic Fishermen's Union; Samuel Somers, Seafood Workers' Union; Alfred White, formerly of Seafood Workers' Union. In New Bedford, Captain John Murley; Captain Daniel F. Mullins; George Feener and Leo Barrett, Atlantic Fishermen's Union; Joseph Sylvia, Seafood Workers' Union; John Linehan, Seafood Producers Association. In Portland, George Ratcliffe, Portland Fish Co.; Russell Yelton, Mid-Central Fish Co.; John McIntosh, formerly of Atlantic Fishermen's Union.

Others include, from the U.S. Fish and Wildlife Service, A. W. Anderson, Director of Division of Commercial Fisheries; B. E. Lindgren, J. J. O'Brien, Joseph Pileggi, Dr. R. A. Kahn, W. H. Dumont, Joseph F. Puncochar, Boris Knake, Homer Haberland, Richard Marchant, W. F. Royce, Howard Schuck, Ray Buller, George Snow, T. J. Risoli, D. W. McKown, C. E. Peterson, E. A. Power, R. A. Nesbit, George Clarke, A. A. Albano, Lee Roane, Paul Paradis, the late Francis Randlett, and Ernest Premetz; Director F. W. Sargent of the Massachusetts Division of Marine Fisheries; R. E. Reed, former Commissioner, Maine Sea and Shore Fisheries; W. D. Heydecker, Atlantic States Marine Fisheries Commission; George Fingold, former Assistant and presently Attorney General, Commonwealth of Massachusetts; John B. Bindloss, Stonington, Conn.; Charles E. Jackson, National Fisheries Institute; Ray Steele, former Counsel, National Fisheries Institute; Ray Kelley, staff economist, Committee of New England.

Gratitude is due, too, to Mrs. Ruth Whitman who edited the manuscript and John Chipman of Harvard University who read critically the theoretical supplement. Mr. Clive Planta, former Secretary-Manager of the Fisheries Council of Canada, favored the author with a forthright statement of Canadian views on some of the New England industry's problems. Though they are not named here, the countless fishermen with whom the author has talked in his travels have his deepest thanks. So, too, do the author's colleagues at Boston College for their interest. The Reverend W. Seavey Joyce, S.J., Chairman of the Boston College Economics Department, was particularly helpful. Professors D. V. Brown, C. A. Myers, and G. P. Schultz of Massachusetts Institute of Technology kindly afforded the opportunity to discuss the antitrust portion of the work before their course in Public Policy. Last, but by no means, least, I wish to thank my wife for her constant encouragement and assistance.

The material employed in the study runs for the most part up to 1952 when the study was finished. I have added a postscript after the last chapter to bring the story more nearly up to date.

DONALD J. WHITE

Boston College, 1953

NORTH ATLANTIC FISHING GROUNDS

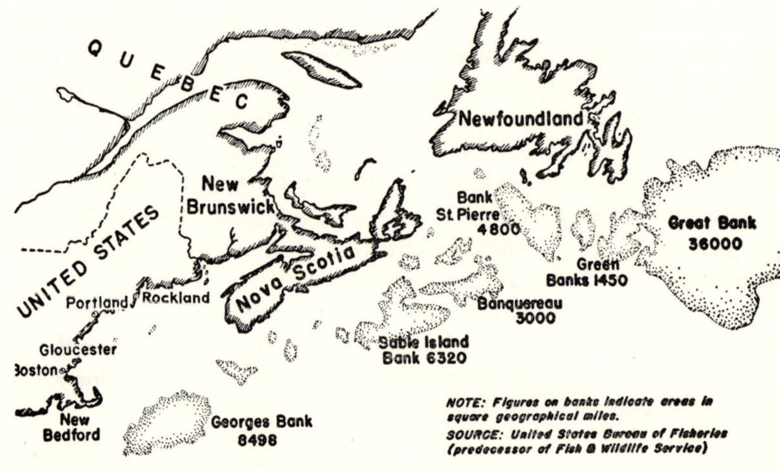

CHAPTER I

THE CHALLENGE

FOR THREE CENTURIES the history of the New England fishing industry has been one of the most exciting chapters in American commerce. From New Bedford in the eighteenth century whaling ships sailed on long perilous journeys into the high seas, and in fact continued to do so until the discovery of oil in Pennsylvania curtailed whale oil markets in the latter half of the nineteenth century. In the 1660's Boston and Gloucester were already providing codfish to the world, thereby furnishing the base for New England's growing ocean trade. And now the same Northwest Atlantic bank fisheries are still being harvested by fishermen from these ports, as well as from Portland and Rockland, Maine. But in these three hundred years the life of the fisherman has changed most radically, as indeed has the very nature of the industry. No longer is the fisherman an isolated hero venturing in his dory from a mother ship into a hostile ocean with doubtful gear, at the mercy of the elements, and in some question of returning alive. Nor is he now so much at the mercy of arbitrary buying conditions, selfish masters, and little pay, even if he does return alive. But if fishing is safer now, and the life of the fisherman somewhat less difficult, there are still conditions which are hazards to the industry and threaten its existence. These problems concern procuring the fish, as well as selling it, the bargaining between fishermen, vessel owners, and dealers, the quality and quantity of the fish available, and the competition with foreign sources.

It is true that Boston, Gloucester, and New Bedford, Massachusetts, and Portland and Rockland, Maine, turn out nearly all the area's packaged fish output, making New England the leading supplier in the nation. In 1951, 80 per cent of the country's domestically produced fresh and frozen packaged fish, amounting to 165 million pounds, flowed from New England.[1] Fishermen sailing from these principal ports land annually about two-thirds of the weight and five-sixths of the value of all fish and sea scallops brought into New England's shores, whereby the region furnishes about 20 per cent of the volume and value of the annual United States and Alaska catch.[2] And yet the leadership enjoyed by New England's fishing centers is far from secure. Four main problems cloud its future:

Profit-Sharing Controversies

New England fishermen do not receive regular individual wages. Alone of all workers in American commerce, fishermen generally work under a system of wage payment called the wage *lay*. In New England it gives them a direct share with vessel owners in the value of the catch and the expenses of each fishing trip, making every man in a sense a partner in the risks or success of each venture.[3] The fishermen are not paid until the fish is sold at the pier and they do not know what their earnings will be until the money is collected.

This profit-sharing arrangement has had a stormy career in Boston, Gloucester, and New Bedford, where, since the late thirties, the Atlantic Fishermen's Union has acted as the militant bargaining agent of virtually all offshore fishermen in insisting upon "fair wages" under the wage *lay*. Rockland and Portland present a somewhat different picture. While the *lay* in those ports conforms generally to that in the Massachusetts centers, Rockland and Portland have had much less conflict, for the union has achieved no contracts and only desultory organization. This is partly because the union has not pressed hard to recruit members, since the Maine ports are less important than those in Massachusetts, as Table 1 shows clearly, and partly because the

---

[1] *Packaged Fish, 1951*, U.S. Department of the Interior, Fish and Wildlife Service, Washington, D.C.

[2] The U.S. Fish and Wildlife Service reports catch and value data annually in the following publications: *Massachusetts Landings by Ports*; *Maine Landings* (published in coöperation with Maine Department of Sea and Shore Fisheries); *New England Fisheries*; and *Fisheries of the United States and Alaska*. The Service also publishes data showing the catch by gear used and area where caught.

[3] The "broker" guarantee constitutes somewhat of an exception to this, however. Please see Chapter V.

union has found it rough going to interest the Maine fishermen in effective organization. (Please note, however, our Postscript.)

TABLE 1

NATIONAL RANK OF LEADING NEW ENGLAND PORTS, 1951 [a]

Port	Landings (millions of pounds)	National rank	Value of landings (millions of dollars)	National rank
Gloucester	260	2	12.6	4
Boston	171	3	14.3	3
New Bedford	79.3	8	11.9	5
Portland	56.9 [b]	.. [c]	2.4 [b]	.. [c]
Rockland	52.1 [b]	.. [c]	2.3 [b]	.. [c]

[a] Source: Ports were ranked according to data reported in C. E. Peterson, "Preliminary Review of the Fisheries of the United States, 1951," *Fishery Leaflet 393*, U.S. Fish and Wildlife Service, Washington, D.C., revised, March, 1952. Landings figures, however, were taken from "Massachusetts Landings — 1951, by Ports" and "Maine Landings, 1951, by Counties." The former is published by the same source as the Peterson data. The latter is published by the Fish and Wildlife in conjunction with the State of Maine Sea and Shore Fisheries.

[b] Landings for Portland are as reported by Fish and Wildlife. Adjusted value of Portland landings and data for Rockland are based on "Maine Landings, 1951, by counties." Portland and Rockland are the most important centers in Cumberland and Knox Counties respectively. Data for sea herring (not included in this study) are excluded insofar as possible from the figures.

[c] Rank not available, but probably not among the first ten ports.

Under the *lay*, price determination clashes have been the central difficulty. The Fishermen's Union, seeking to better its members' incomes, has fought bitterly with the ports' fish dealers for more satisfactory landing prices to boost catch values. Both groups have insisted on different price determination practices, each claiming his own to be "fairer" and more favorable to the industry.

The price disputes have affected collective bargaining directly only where boats are dealer-owned. In Gloucester and New Bedford most vessels are owned by nondealer independents. The union has had comparatively little trouble with these independents over shares in the catch under the *lay*, but it has warred intermittently with the dealers over prices. The union has installed its own auction systems for sales in these ports. The dealers bid for fish in these selling rooms, but many of them have insisted on cutting prices when the fish are delivered to their docks, claiming then that the fish is of low quality and that there is excessive ice and trash in the haul. These factors are not recognized formally in the selling rooms and are sharply disputed by the fishermen at dealers' docks.

Boston has experienced the worst turmoil because most of the port's trawlers are dealer-owned. The fishermen and dealers have clashed both as buyers and sellers and as employers and employees. As the latter, the parties have quarreled continually over selling methods and practices on the New England Fish Exchange. The fishermen insist that this auction market established by the dealers in 1908 often does not yield "fair" prices, even though it operates technically under the terms of a 1919 federal antitrust decree. Collective bargaining between the groups has been a long struggle. The three agreements reached in thirteen years cost the port aggregate losses in fishing time of more than thirteen months.

The persistent price disagreements between the union and the major port dealers have impelled the union to adopt restrictive policies. It has tightened its admission requirements to regulate the supply of union fishermen. Union members have periodically voted catch limitations to enforce their price demands. The effects of the price determination disputes and union policy reprisals cannot be measured with precision, but both probably have tended to limit the industry's competitive ability because they have raised costs and stood in the way of coöperative arrangements to improve the demand for fish.

DECLINE IN FISH POPULATION

The key species of fish in the New England catch have become more difficult to secure in recent years. The operation of market forces has led New England interests to exploit recklessly the limited self-renewing stocks of these species on New England banks and to join with foreign fishermen in "mining" relentlessly the banks in northwest Atlantic international waters. Accumulating evidence indicates that at least some oceanic fish stocks are not exhaustible; heavy fishing pressure tends to have a depleting effect.

Most serious for New England is the scarcity of haddock, Boston's specialty and the area's most valuable groundfish; [4] redfish, the main-

---

[4] "Groundfish" are fish which live on or near the sea floor and may be found in all seasons. They include principally haddock, cod, redfish, hake, pollock, whiting, flounder, and cusk.

stay of the frozen fish industry at Gloucester and Portland;[5] and yellowtail flounder, a highly valuable specialty of New Bedford. Limited supply of these fish means substantially higher unit costs of production as the industry's output per day of fishing effort declines. Shrinking supplies can be obtained only by fishing more intensively in New England waters or by making much longer trips to more distant banks, thereby increasing costs.

MARKETING DIFFICULTIES

New England interests encounter formidable marketing obstacles. The overwhelming preference of consumers for meat severely restricts the demand for fish. In 1949, for example, civilian per capita consumption of meat in the United States was 146 pounds; for fresh and frozen fish it was only 6.5 pounds. Next, consumer preference favors but a handful of the edible species which New England might market. Haddock, redfish, cod, and yellowtail flounder landings in 1950 accounted for three-fifths of the 484.8 million pounds of fish landed at Boston, Gloucester, and New Bedford.[6] Lack of demand is a principal reason for underexploitation of many other species.

Moreover, the available evidence indicates that consumer demand in the past has run strongest on one day of the week and during the Lenten season, chiefly because of the operation of religious factors and custom. Though this feature of fish marketing is diminishing in importance as production of frozen rather than fresh fish increases and the New England product increasingly sells in a national market, custom dictated by religious influences nevertheless remains important, particularly in the high consumption area of the northeast United States. Finally, the distribution system for fish has tended to absorb a high percentage of the retail price. New England interests have stated that wholesale markups of 30 to 40 per cent and retail margins of 50 to 100 per cent have not been unusual. This allocation has held the industry's (and particularly the fishermen's) share of the retail sales dollar to a minimum, even though margins as high as those cited no longer prevail generally.

[5] "Redfish" are known also as "ocean perch" and "rosefish." Redfish are also the principal Rockland product. See Chapter III.

[6] Data from *Massachusetts Landings — 1950, By Ports*, U.S. Fish and Wildlife Service, Washington, D.C.

Foreign Competition

Imports, particularly from the Canadian Maritime Provinces, but on a growing scale from Iceland and other northern nations, jeopardize the sale of New England's packaged fish in the United States market. Imports of groundfish fillets — New England's specialty — increased 830 per cent between 1939 and 1951.[7] Foreign sources supplied about one-third of the American market in 1951; in 1939 they captured about one-twelfth of it.

Natural advantage accounts in part for the growing importance of foreign competitors. They are closer to their fishing grounds and can land fresher, higher quality fish. Moreover, their economies rely heavily on fishing; their workers have few alternative employment opportunities. Ex-vessel fish prices are substantially lower than those in New England;[8] the fishermen's share in the lay is often smaller; labor costs in processing are appreciably lower. Some of the larger New England firms have established plants in foreign areas in order to benefit by these important cost advantages.

Nearly all foreign fillet producers have plans for expanding their sales in the United States. Their governments in many cases are fostering programs to help them toward this goal. Furthermore, price inflation in this country has scaled down the American tariff barrier. Between 1939 and 1948, for example, rising fish prices cut the relation of import duty to import prices of foreign fillets from 32.9 to 12.2 per cent.[9]

The problems of fish scarcity, marketing, and foreign competition challenge the industry to cut costs and broaden its markets. Profit-sharing controversies and related restrictive union policies have tended to hamper such developments. In addition, they have perhaps contributed to the industry's cost handicap. The special challenge to the Fishermen's Union and dealers is to discover a basis for settling their

---

[7] Cf. "Summary of Information Obtained During the Investigation on Groundfish Fillets," *Report on Escape Clause Investigation under Section 7 of the Trade Agreements Extension Act of 1951* (Washington, D.C.: U.S. Tariff Commission, September, 1952), page 82.

[8] "Ex-vessel prices" are those paid to the fishermen.

[9] Cf. *Problems of the Fishing Industry*, Hearings, Subcommittee on Fisheries and Wildlife, Committee on Merchant Marine and Fisheries, U.S. House of Representatives, 81st Congress, 1st Session, February 15 and 16, 1949, page 20.

internal differences and for coöperating in solving their common problems.

The profit-sharing lay links the fishermen's fortunes directly to the production and marketing performance of the industry. Is this a poor arrangement which should be scrapped in favor of a more "orthodox" system of wage payment? If so, why has the lay method prevailed for so long a period? If the lay should be retained, how might its performance be improved? How do Fishermen's Union policies figure in these matters? What kind of forward-looking program might the industry adopt to improve its competitive position? What might be the role of the union? Finally, what are the prospects for adoption of a positive course of action?

Subsequent chapters, we hope, will suggest answers to these questions.

CHAPTER II

THE MODERN INDUSTRY

New England's fishing industry springs from a natural resource that envelopes 260,000 square miles of the world's better fishing territory.[1] Its modern fresh and frozen fish business is the mixed product of stubborn traditional practices and persistent innovations introduced irregularly over the last sixty years.

What are the chief features of modern New England fishing operations? How have technical advances in refrigeration, processing, and transportation helped to mold the present day industry? What obsolete and inefficient practices handicap it? The discussion below covers events up to 1951. Changes since are discussed in Chapter VIII and the Postscript.

Grounds and Species

New England's rambling coastline parallels the southern end of a continental shelf which extends for one thousand miles along the Northwest Atlantic coast from Newfoundland to Long Island.[2] Some eighty species of edible fish and shellfish inhabit the relatively shallow waters within three days sailing distance of the principal ports.[3] About eight or nine have made up the bulk of the main ports' catch. Most important as a group are haddock, redfish, flounder, cod, whiting, pollock, and hake. They are the chief groundfish or bottom-feeding species available on the fishing banks in all seasons. Groundfish are the most prized marine resource of New England and the third

[1] Cf. *Fishery Resources of the United States*, Senate Document No. 51, 79th Congress, 1st Session, March, 1945, page 48.

[2] *Ibid.* pp. 47 and 48.

[3] Cf. Rachel Carson, *Food from the Sea*, U.S. Fish and Wildlife Service Conservation Bulletin No. 33 (Washington, D.C., 1943), pp. 1 and 2.

most productice and valuable in the United States.[4] Several foreign nations, including principally Canada, France, Portugal, and Spain, also draw on the resource and collectively take more than the New England catch.[5]

Mackerel and sea scallops are the other most popular species. The mackerel is a surface-swimming, migratory fish, characterized by as yet unexplained large fluctuations in abundance.[6] The mackerel catch commonly varies by several million pounds from year to year. From late spring to early fall seiners purse in the mackerel schools with gigantic "purse seine" nets. Gloucester has been the home of the "mackerel fleet" since about 1840.[7] Sea scallops are shellfish found on sandy and rocky bottoms at depths of two to 150 fathoms along the New England and Canadian coast. Fishing vessels equipped with dredges scoop them from the sea.[8]

CHIEF FISHING METHODS

Groundfish are swept from the ocean depths with huge, conical nets dragged over the fishing banks by large steel trawlers, operating principally from Boston,[9] and smaller wooden draggers sailing mainly from Gloucester, New Bedford, and Portland. Just after the turn of the century this process of "otter trawling" began to displace fishing by hand lines from open dories which were carried to the fishing grounds by sail-powered schooners. The dories, manned by one or two men, were expected to bring back a certain amount of the haul, but the dangers of independent fishing made the outcome far more uncertain than it is now.[10]

[4] Cf. *Fishery Resources of the United States*, Senate Document No. 51, 79th Congress, 1st Session, March, 1945, page 48.

[5] *Ibid.*, page 47.

[6] Cf. O. E. Sette, *Biology of the Atlantic Mackerel of North America*, part I, "Early Life History," Fishery Bulletin No. 38, U.S. Dept. of the Interior, Fish and Wildlife Service, 1943.

[7] Cf. R. McFarland, *A History of the New England Fisheries* (Univ. of Pennsylvania, 1911), page 267.

[8] Cf. *Fishery Resources of the United States*, Senate Document No. 51, 79th Congress, 1st Session, March, 1945, page 95.

[9] For a discussion of Boston trawling see T. D. Rice, "Fishing — Out of Boston," a booklet available from the Massachusetts Fisheries Association, Fish Pier, Boston, Massachusetts.

[10] "Otter trawling" has been called also "beam trawling." Both names derive from methods used to hold the mouth of the net open below the ocean surface.

The *Spray*, New England's first successful otter trawler, was built in 1905 for John R. Neal, a Boston fish dealer.[11] The vessel incorporated most of the features of similar craft operating from English ports as early as the 1870's. Nearly a year and a half passed before the *Spray* became profitable. It then proved so worthwhile, however, that Neal's firm, the Bay State Fishing Company, acquired nine trawlers by 1915 and became the biggest fish-producing company in New England.

TABLE 2

AVERAGE CATCH PER TRIP BY LINE VESSELS AND BY OTTER TRAWLERS ON GEORGES BANK AND SOUTH CHANNEL, 1910 TO 1914

(thousands of pounds)

	1910	1911	1912	1913	1914
Line vessels	32	39	36	28.5	27.0
Otter trawlers	50	45.5	51	44	43.3

Source: *Otter-Trawl Fishery*, House Document No. 1519, 63rd Congress, 3rd Session, January 25, 1915, pp. 39–40. Figures are average totals for haddock, cod, and hake.

Otter trawling did not make its debut without opposition. Hook-and-line fishermen and schooner-owners besieged Congress to prohibit the importing and landing of trawler-caught fish.[12] They charged that the method would deplete the fishing grounds and destroy their gear.[13] Their plea failed. Studies by the United States Commissioner of Fisheries discounted the opponents' claims. Looking to the future, however, the Commissioner's report warned that the new otter trawl method did destroy large quantities of commercial fish too small to

[11] Cf. *Otter-Trawl Fishery*, House Document No. 1519, House of Representatives, 63rd Congress, 3rd Session, January 25, 1915, page 15. Also, M. F. Blanchard, "Boston's Place in Our Great Fishing Industry," *The Fishing Gazette*, XXXII, no. 9 (New York, February 27, 1915), pp. 292 and 312; and "Trawling and Dragging in New England Waters," *The Atlantic Fisherman's Almanac, 1929*, Atlantic Fisherman, Inc. (Boston, Mass.), pp. 88–96. An earlier otter trawler, the *Resolute*, operated from Gloucester in the late nineteenth century, but did not prove sufficiently successful to start widespread adoption of the method.

[12] Cf. *Prohibiting the Importing and Landing of Fish Caught by Beam Trawlers*, Hearings, Committee on Merchant Marine and Fisheries, House of Representatives, May 28 and 29, 1912.

[13] Cf. "The Menace of the Otter-Trawl," *Boston Transcript*, April 19, 1913. The charge has persisted. See L. Allen, "Fish Industry Facing Crisis," *Boston Sunday Post*, January 23, 1949.

market because the net, when dragged over the ground, tended to sweep up everything in its way. It pointed out that unregulated use of the method might in the future cause the fishery resource to suffer from overfishing and recommended that the use of the otter trawl be restricted to designated banks and grounds.[14]

The recommendation went unheeded, because such regulation, to be fair and effective, would have required the coöperation of other nations, principally Canada, Newfoundland, and France, which also drew on the Northwest Atlantic fishery resource. The New England industry shifted rapidly to otter trawling, for the method tremendously increased the catch per trip and also eliminated the dangerous practice of solitary hook-and-line trawling from dories on the banks miles from shore. By 1925 there were more than 100 boats and vessels operating trawling gear in New England waters.[15] By that year, smaller boat fishermen had developed the "baby trawler," or dragger, to complement the larger steel vessels.[16]

Modern trawlers and draggers boast countless improvements over early models. Diesel power has displaced coal-burning engines, opening up more space for fish storage and crew and eliminating firemen as crew members. Mechanical refrigeration supplements ice in up-to-date aluminum, cork, or stainless steel storage rooms. The trawlers acquired radios in the twenties; today nearly all trawlers and draggers have radio-telephones. They have loran for direction finding and the fathometer for undersea sounding. A few vessels began using radar during 1952. These instruments remove much of the danger and guesswork from fishing operations and keep the ship in constant touch with marketing conditions on land. The Vigneron-Dahl trawl, introduced from Europe during the 1920's, has greatly improved trawling efficiency. It enlarges the effective catch area of the deep-sea fishing net by placing the large wooden "doors" which hold the net mouth open some distance forward of the net rather than adjacent to it.

Today's large steel trawlers have an "iced down" fish capacity of

[14] Cf. *Otter-Trawl Fishery*, House Document No. 1519, 63rd Congress, 3rd Session, January 25, 1915, pp. 92 and 98.

[15] Cf. "Trawling and Dragging in New England and Waters," *The Atlantic Fisherman's Almanac, 1929*, Atlantic Fisherman, Inc. (Boston, Mass.), pp. 88–96.

[16] *Ibid*. Captain Daniel F. Mullins of New Bedford pioneered in this development with extensive experimentation in the New Bedford flounder fishery.

250 to 300 thousand pounds and carry crews of about 17 men. The "baby trawlers" are most often wooden and, on the average, have from one-third to one-half the capacity of the larger steel vessels, carrying crews which usually range from nine to twelve in number.

REFRIGERATION

Progress in refrigeration after 1900 opened up to the New England industry valuable new marketing possibilities for its very perishable commodity. In the old days most of the fish was salted on board and only a small per cent was brought home fresh. Successive improvements have enabled fresh and frozen fish sales to supplant the waning salt fish business.

Cold storage facilities lengthened the period of consumption and widened the market. Between 1900 and 1919 the number of freezers in Massachusetts rose from four to twenty-two; between 1913 and 1918 the amount of fish frozen and stored annually increased from 17.8 million pounds to 51.9 million pounds.[17] At that time the fish were frozen whole in pans after immersion in water to give their skins a protective glazing to prevent dehydration. This process allowed storage for a period of up to twelve months. It preserved nutritive value, but it cut palatibility so that consumers naturally considered the frozen product inferior to fresh fish.

The introduction of "quick freezing" techniques in the twenties repaired this deficiency. The United States Bureau of Fisheries imported from Europe in 1918 the first "quick freezing" machine for experimental purposes;[18] a few years later Clarence Birdseye of Gloucester and Dr. Harden F. Taylor of the Bureau of Fisheries developed independently separate processes which had the greatest effect in New England. Birdseye's process became the property of General Seafoods;[19] Taylor's method was adopted by the Atlantic Coast

---

[17] Cf. *Report of the Joint Special Recess Committee to Investigate the Fishing Industry*, House No. 1725, Commonwealth of Massachusetts, April, 1919, pp. 49–50.

[18] This was the Ottesen Brine Freezer. See J. M. Lemon, *Developments In Refrigeration of Fish in the United States*, Investigational Report No. 16, United States Dept. of Commerce, Bureau of Fisheries (1932), page 3.

[19] Cf. "Let Them Eat Cake — the Story of General Foods," *Fortune*, October, 1934, page 135. The name General Seafoods has been superseded by the name Birdseye Division, General Foods Corporation.

Fisheries Company;[20] these concerns were two of the largest operating in New England. Both techniques consisted basically in sharp freezing of fish at temperatures as low as 60° below zero Fahrenheit.

TABLE 3

UNITED STATES REFRIGERATED STORAGE CAPACITY DEVOTED TO FISH,
OCTOBER 1, 1949

| | | Capacity (thousands of cubic feet) | |
Region or state	No. of firms	Freezer 29°F and below	Cooler above 29°F through 50°F
By region:			
New England	29	365	2,665
Middle Atlantic	53	824	2,874
E. No. Central	47	257	2,268
W. No. Central	24	69	748
South Atlantic	32	205	903
E. So. Central	15	22	374
W. So. Central	22	9	537
Mountain	9	10	66
Pacific	43	795	15,266
U.S. Total	274	2,556	25,701
New England region, by states:			
Maine	6	37	227
Massachusetts	18	326	2,360
Other States	5	2	78
New Eng. Total	29	365	2,665

Source: *A Survey of the Capacity of Cold Storage Warehouses in the United States As of October 1, 1949,* U.S. Dept. of Agriculture, as reported in Fishery Products Report No. 231, U.S. Dept. of the Interior, Fish and Wildlife Service, *Boston Market News,* November 29, 1950, page 4.

The rapidity of freezing preserved a very high degree of fresh flavor in the frozen product, for it allowed the ice crystals to produce less rupture in the food cells.

The development of these systems stirred cold storage executives to improve cold storage warehouse freezing. Freezing time was cut from three days to a few hours by lowering temperatures, using smaller freezing rooms, and circulating cold air rapidly with large fans. The consequent improvement in the warehouse product helped it rival closely the strictly "quick frozen" variety. Table 3 shows the

[20] Taylor became the president of Atlantic Coast Fisheries Co. See J. C. Wheeler, "A Review of Massachusetts Fresh and Frozen Fishery Products"

extent of refrigeration facilities in the United States utilized by fishing interests in October, 1949.

Processing

The greatest stimulant to improved marketing blossomed in Boston in the fall of 1921. This was the introduction of filleting at the port.[21] Dana F. Ward, a prominent Boston fish dealer, initiated the development. Retail dealers had for many years cut the boneless meat sections from the sides of fish in preparing fish for their customers. To broaden his market, Mr. Ward cut such "fillets" from haddock; he wrapped the fillets in parchment paper and shipped them uniced to Washington, D.C. where they met with a most favorable reception. The business expanded rapidly as John C. Wheeler, manager of the Bay State Fishing Company, devised a plan for refrigerating the fillets in shipment. Until 1927, haddock comprised nearly all the industry's fillet output as the species seemed to meet every requirement of quality, size, abundance, and price for processing. After that date more than a dozen other species gradually appeared on processing lists.

Filleting had great repercussions in the main ports. It gave birth to "branded" packaged fish marketing and thus gave the dealers an opportunity to compete on a differentiated product basis.[22] It led to the greater participation of chain stores in the fresh and frozen fish business.[23] It made Boston the nation's undisputed haddock capital,

---

(Mimeographed), Walpole, Mass., 1942. Mr. Wheeler was an important figure in the industry from 1916 to 1946. See also, H. F. Taylor, "A New Method of Brine Freezing for Fish," *Canadian Fisherman*, March, 1925.

[21] Information on the introduction of filleting was obtained from the following sources: R. H. Fiedler, *Trade in Fresh and Frozen Packaged Fish Products*, Economic Circular No. 63, U.S. Dept. of Commerce, Bureau of Fisheries, Washington, D.C., May, 1928; J. C. Wheeler, "A Review of Massachusetts Fresh and Frozen Fishery Products" (Mimeographed), Walpole, Mass., 1942; interviews with Mr. Wheeler; the private correspondence of Mr. Wheeler and Mr. Ward.

[22] Filleting made possible "monopolistic competition" among the dealers. See E. H. Chamberlin, *The Theory of Monopolistic Competition* (Cambridge, Mass., Harvard University Press, 1948).

[23] Mr. Joseph O'Brien, Manager of the National Fish Division, A & P Tea Co. stated that the latter chain began buying fish on the Boston Pier around 1926; he was the fish buyer for the concern.

lifting the port out of the acute depression it suffered as an aftermath of the first World War.

TABLE 4

PRODUCTION OF FRESH AND FROZEN PACKAGED FISH IN THE UNITED STATES AND IN NEW ENGLAND, 1929–1951, AND NEW ENGLAND PRODUCTION AS A PERCENTAGE OF UNITED STATES PRODUCTION

Year	U.S. production (millions of pounds)	New England production	
		(millions of pounds)	Percentage of U.S. output
1929	84.4	73.5	87
1930	80.0	68.5	85
1931	65.6	54.3	82
1932	52.0	41.7	80
1933	58.9	48.9	83
1934	73.6	59.6	81
1935	107.1	86.6	80
1936	121.5	103.5	85
1937	113.4	92.8	82
1938	117.2	98.8	84
1939	131.3	111.1	84
1940	126.6	107.7	85
1941	176.5	152.6	86
1942	172.9	144.4	83
1943	158.6	122.0	76
1944	172.3	139.3	86
1945	204.9	160.6	78
1946	188.5	152.9	81
1947	185.0	156.6	84
1948	193.5	155.5	80
1949	194.0	155.3	80
1950	191.5	150.5	78
1951	205.5	164.7	80

Source: U.S. Fish and Wildlife Service, *Packaged Fish Reports*. Production figures are rounded to nearest tenth. The 1929–1931 figures include a small amount of smoked fish.

The waste from filleting made possible a sizeable by-products development — the production of fish meal, oil and solubles, animal food, and fertilizers. The elimination of waste — about 60 per cent of total fish weight — cut out freight charges for bones, head, and tail, and broadened the industry's market area. Fillets proved well suited to freezing; dealers shipped carload lots to all sections of the country where they were stored in public or private cold storage warehouses for redistribution.

TRANSPORTATION

Progress in transportation has joined with refrigeration and filleting to help the industry better its products and extend its markets. The development of refrigerated trucking after 1930 provided a direct, flexible delivery channel which could compete with the railroads.[24] By 1940, about 70 per cent of Boston's fresh and frozen fish shipments traveled by truck.[25] Air freight entered the scene after the second World War.[26] In 1949, eight cities had direct air service on fresh fillets from Boston; one airline was delivering 50,000 pounds weekly.[27]

TABLE 5

DISTRIBUTION OF FRESH AND FROZEN FISH FROM BOSTON, MASSACHUSETTS, 1940

(millions of pounds)

Destination of shipments	Methods of Transportation			
	Truck	*Rail express*	*Rail freight*	*All*
Boston Metropolitan area	50	..	..	50
M.A.[a] to 200 miles	30	10	..	40
201 to 500 miles	18	..	42	60
501 to 1000 miles	32.4	..	3.6	36
Over 1000 miles	22.5	..	7.5	30
Total	152.9	10	53.1	216

Source: A. W. Anderson, "Wartime Transportation and the Fishing Industry," *Fishery Market News*, April Supplement, 1943, U.S. Fish and Wildlife Service, Washington, D.C.
[a] Metropolitan Area.

Transportation advances have melted the obstacle of marketing distance. In 1922, Boston distributed 56 per cent of its catch within Massachusetts.[28] By 1940, 58 per cent of Boston's fresh and frozen

[24] Cf. "The Coming of the Trucks," *Fishing Gazette*, N.Y., June 15, 1932, pp. 107–111.

[25] Cf. A. W. Anderson, "Wartime Transportation and the Fishing Industry," *Fishery Market News*, April Supplement, 1943 (U.S. Fish and Wildlife Service, Washington, D.C.).

[26] Cf. J. P. Blount, "Fish Deliveries by Airplane," *Fishing Gazette*, N.Y., June, 1945, page 90.

[27] Cf. Robert Ramspeck, "Flying Fish to Market," *Yearbook*, National Fisheries Institute, Washington, D.C., 1949.

[28] Cf. L. T. Hopkinson, *Trade in Fresh and Frozen Fishery Products and Related Marketing Considerations in Boston, Mass.*, U.S. Bureau of Fisheries, Washington, D.C., 1922.

fish was moving more than two hundred miles to reach consumers; nearly 15 per cent of it, or thirty million pounds, was traveling to destinations over one thousand miles away.[29] Table 5 shows in detail Boston's fresh and frozen shipments for 1940 by distance and method of transportation.

Obsolete and Inefficient Practices

Despite its modern trappings, the industry clings to many outmoded, inefficient practices. Most outstanding is the age-old method of handling fish aboard the vessel and at the docks. The fish are "pitchforked" from one place to another at least three times before they complete the trip from the ocean to the dealers' shops. Conveyor systems for loading and unloading are a rarity.

United States Fish and Wildlife Service technologists have suggested effective but inexpensive techniques to eliminate "pitchforking" and the deteriorating skin punctures which follow from it.[30] The suggestions have gone unheeded. Even the threat of action by public health authorities in 1940 to prevent possible disease did not wipe out the practice in Boston. The fish handlers refused to unload the vessels without forks, and many of the dealers, fearing the expense of installing new methods, tacitly supported the "lumpers" position.[31]

Another lapse occurs in the removal of fish from the dock to some dealers' shops. Many processors cannot handle supplies as they are received. They often leave the fish in open carts unprotected from the weather for more than half a day. Moreover, the processing shops have no impartial check on quality beyond the minimum specifications of state and federal pure food and drug laws. The Boston dealers sought unsuccessfully to remedy this situation in 1939 when the bill for fish inspection which they had introduced into Congress failed

---

[29] Cf. A. W. Anderson, "Wartime Transportation and the Fishing Industry," *Fishery Market News*, April Supplement, 1943, U.S. Fish and Wildlife Service, Washington, D.C.

[30] Cf. J. F. Puncochar and S. R. Pottinger, "Suggestions for Handling Trawler-Caught Fish," *Commercial Fisheries Review*, December, 1947, U.S. Fish and Wildlife Service, Washington, D.C., pp. 12–18.

[31] "Lumpers" are those unloading fishing vessels. See "Fish Handlers' Strike Averted," *Boston Globe*, January 8, 1940.

of adoption.[32] The dealers have not succeeded in developing a self-policing system of quality inspection through their associations in any of the principal ports. Some shops, particularly the smaller ones, have no effective sanitation controls. These deficiencies handicap the industry in its most vital task: quality control of a highly perishable product.[33]

[32] The Boston dealers coöperated with other eastern seaboard fishing interests in backing the bill, S.2380, introduced into the U.S. Senate on May 9, 1939.

[33] Cf. D. K. Tressler, "Quality Control Vital To Success in Frozen Foods Industry," *Food Industries*, June, 1938, page 320.

CHAPTER III

THE PRINCIPAL FISHING PORTS

THE FISHING CENTER of Boston, of Gloucester, of New Bedford and of Portland has each had its own origin, history, and development. The product specialties and market organization of each are also highly individual. We shall first attempt to clarify the individual characteristics of each port before we discuss their common problems.

BOSTON

The foundation for the twentieth-century expansion of Boston's fresh and frozen fish business was laid in 1914 when the Boston Fish Pier, then the world's largest, was opened for business.[1] The State of Massachusetts built the pier at a cost of over a million dollars under terms of a 1910 agreement with the Boston Fish Market Corporation, a company which the Boston fish dealers formed to acquire a new location for the industry.[2] Their old facilities on "T" Wharf, which they had occupied since 1884, had become unsanitary and inadequate.[3]

Under the agreement with the state, the corporation leased the pier for fifteen years from October 1, 1913. The lease provided for three subsequent extensions of fifteen years until 1973. The corporation spent a million dollars for buildings and rented space to the dealers on an annual basis. In 1913 the Commonwealth Ice and Cold Storage Company was formed to erect and maintain on the pier the nation's

[1] Cf. "Boston As a Fishing Port — Past and Present," *Fishing Gazette*, May 1, 1915, N.Y., pp. 545–552.

[2] All arrangements concerning the agreement are described in Commonwealth of Massachusetts, *Investigation of the Fish Industry*, House No. 1591, May 1918, pp. 11 and 14.

[3] Cf. M. F. Blanchard, "Boston's Place in Our Great Fishing Industry," *Fishing Gazette*, February 27, 1915, page 257.

TABLE 6

Facilities and Employment in the New England Fishing Industry at Boston, Gloucester, New Bedford, Portland, and Rockland, 1950–1952

Port	Vessels			Number of fishermen	Number of processing plants	Number of shore workers	Number of Fish freezing plants
	Small	Med.	Large				
Boston	51	30	48 [a]	1092	48	2000	9
Gloucester	60	114	13	2300	20	2200	8
New Bedford	113	33	3	1350	12	650	2
Portland	52	9	1	200	9	.. [b]	4
Rockland	76	4	1	200	3	.. [b]	2
Total	352	190	66	5142	92	4850	25

Source: Fishing vessel data from *Official Yearbook of the Fishing Masters Association, 1950*, Boston, Mass.; data on fishermen, shoreworkers, and processing plants from the U.S. Fish and Wildlife (estimated) except for Boston. The U.S. Department of Labor, Bureau of Labor Statistics furnished that figure. Processing plant data are from "Producers of Packaged Fish — 1951," *Commercial Fisheries SL-161*, U.S. Fish and Wildlife Service, October, 1952.

[a] As of 1952, about four of these vessels, property of the General Foods Corp. were landing fish in Rockland, Maine.

[b] Reliable data are not available. However, there are probably about 300 in Rockland and something over 350 in Portland.

largest icing and cold storage plant devoted exclusively to the fishing industry. The Boston Fish Market Corporation held the controlling interest in this concern.[4]

These improved facilities combined with the trawling, processing, refrigeration, and transportation advances which we have already discussed [5] enabled Boston landings to soar from less than a hundred million pounds in 1914 to a peak of nearly three hundred and forty million pounds in 1936. From 1914 to 1951 haddock and scrod (small haddock) [6] dominated Boston landings. They constituted half the total weight; their proportion of total value rose from approximately 40 per cent to over 60 per cent through the period.[7]

Nearly all the fish landed at Boston are sold through the New England Fish Exchange at the Boston Fish Pier. The exchange, an auction market, was set up by the dealers in 1908 under their own rules, limiting the buying privileges to those concerns doing business at "T" Wharf when the exchange was formed.[8]

These firms banded together into two competing holding companies by 1916. The Bay State Fishing Company, which controlled a major share of the port's landings because it operated all of the largest trawlers, absorbed eight of the other dealers.[9] The remaining concerns formed the Boston Fish Pier Company in an attempt to match Bay State's strength in the market.[10] Both combines made enormous profits during the first World War.[11]

In 1918 the state of Massachusetts and the United States government prosecuted this duopoly,[12] the Boston Fish Market Corporation

---

[4] Cf. Commonwealth of Massachusetts, *Investigation of the Fish Industry*, House No. 1591, May 1918, pp. 20 and 21.

[5] See Chapter II.

[6] In the industry "scrod" means "small." It is applied to both haddock and cod; when used alone, however, it usually designates small haddock ($1\frac{1}{2}$ to $2\frac{1}{2}$ pounds in weight). See testimony of E. L. Dunn, Manager of the New England Fish Exchange, in *Commonwealth of Mass. v. Patrick J. McHugh et al.*, Equity No. 58644, Suffolk Superior Court, Boston, Mass., June, 1947, page 773.

[7] Percentages are computed from landings data furnished by U.S. Fish and Wildlife Service.

[8] Cf. Commonwealth of Massachusetts, *Investigation of the Fish Industry*, House No. 1591, May 1918, page 12.

[9] *Ibid.* pp. 23–26.

[10] *Ibid.* pp. 18 and 19.

[11] *Ibid.* pp. 26–29.

[12] The Bay State Fishing Company and the Boston Fish Pier Company con-

TABLE 7

BOSTON LANDINGS, 1914–1951

Year	Haddock [a]		Scrod [b]		All species	
	Millions of pounds	Millions of dollars	Millions of pounds	Millions of dollars	Millions of pounds	Millions of dollars
1914	40.99	1.09	6.24	0.107	92.23	2.61
1915	36.04	1.01	11.80	.17	97.40	2.89
1916	34.35	1.22	14.20	.31	98.26	3.70
1917	34.04	1.79	11.47	.40	98.16	5.12
1918	46.14	2.45	6.05	.23	109.23	6.56
1919	55.56	2.13	2.30	.06	103.21	4.70
1920	64.72	2.48	...	...	118.30	6.11
1921	55.22	1.73	...	...	104.28	4.18
1922	52.67	1.50	0.20	.004	106.03	4.01
1923	56.12	2.05	4.42	.09	123.96	5.42
1924	57.12	1.85	11.02	.19	130.63	5.39
1925	61.39	2.10	12.89	.28	148.72	6.08
1926	71.46	2.59	10.41	.24	167.06	6.99
1927	96.12	2.88	13.74	.27	194.88	7.37
1928	124.79	4.30	12.05	.27	218.35	8.81
1929	151.59	5.79	9.11	.22	255.62	10.73
1930	159.28	5.84	7.82	.16	285.21	10.87
1931	106.03	3.82	14.26	.26	219.93	7.91
1932	87.08	2.37	27.65	.40	215.53	5.37
1933	90.35	2.57	32.53	.64	232.51	6.09
1934	71.85	2.39	53.81	.14	243.60	6.73
1935	100.64	2.69	47.15	1.00	307.37	7.73
1936	97.59	3.31	38.17	0.94	339.16	9.58
1937	96.82	2.89	33.10	.76	324.59	8.47
1938	85.43	2.46	43.02	.71	318.73	7.62
1939	79.46	2.43	49.77	1.10	295.35	8.01
1940	73.30	2.97	36.87	1.19	252.77	8.57
1941	80.05	3.57	59.00	2.24	299.33	11.52
1942	54.39	3.97	45.81	2.94	194.65	12.55
1943	42.70	4.21	32.15	2.80	142.97	12.60
1944	59.84	4.64	13.16	0.94	151.76	10.59
1945	59.30	4.70	6.99	.53	188.16	13.79
1946	57.78	6.08	13.64	1.31	158.15	13.69
1947	70.25	6.13	37.22	2.42	202.66	14.98
1948	58.13	5.79	47.18	3.98	199.98	16.18
1949	52.85	4.66	37.30	2.57	172.47	12.19
1950	47.84	5.12	59.54	4.06	172.03	13.56
1951	46.47	4.92	60.45	4.59	171.02	14.31

Source: U.S. Fish and Wildlife Service.

[a] These data apply to haddock weighing over $2\frac{1}{2}$ pounds. (Figures rounded in most instances to nearest hundredth.) Negligible salt-fish figures excluded.

[b] These data apply to haddock weighing up to $2\frac{1}{2}$ pounds. (Figures rounded in most instances to nearest hundredth.) Landings of scrod were negligible in 1920 and 1921.

and the New England Fish Exchange, for promoting monopoly and restraining trade. In the state action seventeen prominent fish merchants were found guilty and received fines and prison sentences.[13]

The federal trial, based upon the Sherman and Clayton Anti-Trust Acts, produced a decree which revamped the Boston industry.[14] It wiped out the two holding companies, dissolving the Boston Fish Pier Company and divesting the holdings of the Bay State Fishing Company of eight other dealerships. It set limits to the size a Boston fish firm might attain; it forbade the dealers to bid, directly or indirectly, on fish caught by their own vessels. The decree restrained permanently the Boston Fish Market Corporation from discriminating against any dealer in leasing space on Boston Fish Pier. It ordered the New England Fish Exchange to extend buying privileges to any wholesale or retail dealer on terms as favorable as those enjoyed by Exchange members. The Exchange was to refrain from trading in fish on its own account and was to submit for court approval new rules governing buyers and sellers.

Since 1919 the Exchange has operated under the court-approved rules.[15] Commission men act as agents of the vessel owners during the fish auctions. Each species on each vessel is sold separately. Buyers have the right to offer a price per hundred pounds at which they will take the entire amount of a species offered for sale from a given boat; or they may offer to buy a "scale" which is a quarterly lot. The latter feature broadens the market by affording the small dealer a chance to bid. The seller decides whether to accept bids on a "scale" basis or to take one buyer's offer for all the species offered. Moreover, the highest price bid takes the fish only if the seller is willing; he may refuse the bid and hold over for a subsequent selling hour. Sales commence each day at 7:45 A.M. and continue until 5 P.M., if necessary. Any fish unsold at that hour await the next day's

---

stituted the duopoly. For discussion of duopoly see E. H. Chamberlin, *The Theory of Monopolistic Competition* (Cambridge, Mass.: Harvard University Press, 1948).

[13] Cf. *Commonwealth v. Dyer*, Massachusetts Superior Court, August Sitting, 1918.

[14] Cf. *U.S. v. the New England Fish Exchange et al.*, Equity Document 810, U.S. District Court of Massachusetts, December 4, 1919.

[15] Cf. New England Fish Exchange, *Rules and Regulations*, Boston, Mass., 1935 (revised edition).

sale. The seller has one other important privilege. He may place an "upset price" on his fish below which no bid will be accepted. The "upset price" provision is an advantage for a dealer owning vessels. It insures that he get at least as much for the fish landed by his boat as those fish would be worth to him for processing purposes. During the second World War, when Office of Price Administration ceilings in effect set the prices for all fish landed, the "upset price" feature resulted (in the sellers' market which prevailed) in the large boat-owning dealers taking virtually all the fish landed. Other dealers were able to secure supplies only by purchasing it from the vessel-owning dealers at the primary market ceiling plus the OPA controlled primary wholesaler margin of two cents per pound.[16]

The Exchange oversees all transactions. All sales must be registered and settled for through the exchange, which holds bonds posted by the dealers to cover their purchases. A buyer may split his purchase with others, but only after he has completed the purchasing agreement and has arranged with the seller all the conditions of sale. Exchange employees expedite the auction process by posting in advance the quantity of each species of fish, together with the name of the bank upon which the fish were caught, under the respective vessels' names on blackboards around the Exchange hall. These listings are known as the vessels' "hails."

A singular feature of the Exchange rules is the "sellover," a provision that there may be a resale, as second or third quality, of any fish which do not qualify as first quality when they are removed from the vessel after the original sale. Unfortunately, the rules do not furnish any objective standards for distinguishing first from second quality fish. They specify merely, "No. 1 fish must be bought as number 1 fish. No. 2 fish must be bought as number 2 fish." [17] Both categories are considered saleable as fresh fish. The rules do single out third-quality fish as unsatisfactory for sale as fresh fish, but the dealers have rarely invoked such a claim during the last ten years. They have, however, insisted often that the fish are "No. 2."[18]

[16] Cf. Federal Trade Commission, *Report of the Federal Trade Commission on Distribution Methods and Costs — Part VIII, Cost of Production and Distribution of Fish in New England* (Washington: U.S. Government Printing Office, 1945), pages 51 and 52.

[17] Cf. The New England Fish Exchange, *Rules and Regulations*, page 18.

[18] The *Rules* do provide for a "sellover" also when there is an "overrun" of

The associated sellovers have provoked constant controversy between the fishermen and the dealers. The sellover nearly always reduces fish prices; the fishermen's incomes are linked directly to prices under the profit-sharing lay. The fishermen, therefore, through the Atlantic Fishermen's Union which has represented them since 1937, have fought the sellover system bitterly to protect their wages.

The clash has carried over into collective bargaining because many of the port's trawlers are owned or managed by the larger fish dealers. The trawlers account for up to 80 per cent of the fish landed annually. The thorny difficulties of price determination and collective bargaining which harass the Boston market will be discussed and analyzed in Chapter V and the Appendix.

The costs and profits of Boston's dealer-boat operators are obscure. These men refused to coöperate with the Federal Trade Commission's inquiry into these matters in 1945;[19] profit and loss information is closely held, or it is published in consolidated statements which show the combined operations of fish dealing and other lines.[20] Several indicators of profitability are available, however.

First, thirty of the larger vessels operating in 1947 were built during the period 1930–1941 when fish prices were extremely low. This indicates that returns were sufficiently high to attract the necessary capital; owners interviewed set the "capital attraction rate" at 10 per cent on invested funds. Moreover, the industry rapidly refitted more than thirty vessels at the end of the second World War to restore them to fishing after service in the war effort. Although the cost of gear and outfitting had trebled since 1941, this did not seem to deter the reconversion. Finally, industry interests moved quickly to sell twelve trawlers to the United States government in 1949 for operation in Germany, confirming the slump in profitability from wartime levels.

Other data support the indicators already cited. During the period of the thirties, stores on the Boston Fish Pier remained in demand.[21] The only major change in the directory of firms operating on the pier

---

fish, that is, when the vessel reports in the "hail" less than was actually caught. In practice, however, "sellovers" for this reason are not an issue.

[19] *Report of the Federal Trade Commission on Distribution Methods and Costs*, Part VIII, June 30, 1945, page 91.

[20] For example, General Seafoods, now Birdseye Division, General Foods.

[21] Information obtained in an interview with Mr. Joseph E. Delay, Assistant Treasurer, the Boston Fish Market Corporation.

resulted from the expansion of the General Seafoods Corporation which bought out the Bay State Fishing Company. The Boston Fish Market Corporation, which rents the pier buildings to the dealers, increased its surplus account from $115,000 in December, 1931, to nearly $188,000 in December, 1939.[22] Examination of balance sheets made public under Massachusetts law reveals that the second World War period was very lucrative for the dealers. Seventeen firms increased their combined surplus accounts from $935,000 in 1941 to $2,200,000 in early 1947. Between that time and early 1949 prosperity waned. The largest firm in the sample added another $500,000 to its surplus account, but the remaining sixteen companies as a group barely maintained their 1947 standing.[23]

TABLE 8

SALES, PROFITS, AND PROFIT RATES FOR THREE MEDIUM BOSTON FIRMS AND TWO GLOUCESTER FIRMS, 1947, 1949, AND 1951 [a]

Item	1947	1949	1951
Net sales	$5,626,181	$5,623,561	$6,233,311
Net operating profits	44,708	(loss) 78,823	55,735
Net worth (end of year)	585,204	730,893	821,502
Ratio (per cent) of profit to net sales	0.8	....	0.9
Ratio (per cent) of profit to net worth	7.6	....	6.8

Source: Compiled from company records by the U.S. Tariff Commission.

[a] Operating profits (before Federal income tax) as used in the above tabulation means profits from filleting and the buying and selling of other fishery products and excludes income and expense from other sources, such as investments, sale of capital assets, etc. Net worth is capital stock plus surplus.

Limited data obtained by the U.S. Tariff Commission and shown in Table 8 indicate that returns improved between 1949 and 1951. The data are for three medium-sized firms in Boston and two in Gloucester. For the five firms combined, profits averaged 6.8 per cent of net worth in 1951, whereas in 1949 four of the five concerns lost money and losses of the group totaled nearly $79,000. While no cost data

[22] Cf. Boston Fish Market Corporation, "Financial *Statement*," December 31, 1931 and December 31, 1939.

[23] Balance sheet analysis does not, of course, completely reveal profitability. Changes in surplus fail to record the distribution of dividends; surplus may increase through asset revaluation as well as through earnings. Nonetheless, the probability is that the surplus changes we have noted understate rather than exaggerate profitability. The dealers would find little advantage in giving the Massachusetts Tax Commissioner an inflated statement of condition.

are available, the indications are that vessel operations were more profitable in 1951 than in 1949. With landings of about one million pounds less than in that year, the gross value of the catch in 1951 ran about 17 per cent higher. Many in the industry were not satisfied with 1951 returns, however. They joined with operators from the other principal ports in the tariff petition discussed in Chapter VI.

GLOUCESTER

In 1933 this world-famed fishing port reached its lowest ebb in production and prosperity.[24] Once the salt fish center of the world, Gloucester's fortunes declined steadily after 1900 as lower-cost foreign competitors gradually priced the port out of the salt fish market. After 1935, however, the Gloucester industry staged a successful comeback. By 1943 it surpassed Boston in volume of landings and led all New England as a producer of fish.[25]

The springboard for this remarkable resurgence of fishing activity was the development of a huge midwestern market for redfish, a species which New England fishermen had discarded as trash prior to 1933.[26] Chance experimentation revealed that the redfish yields a small white fillet with the taste and texture of fresh water perch. Marketed frozen as "ocean perch," the redfish fillet quickly gained favor in the Chicago area, for it suited the taste of Great Lakes fish consumers.

Although the discovery that led to the development of the market for redfish seems to have originated in Boston, Gloucester capitalized on the find.[27] The location of Gloucester enabled it to take advantage

---

[24] In April, 1933, the Gloucester vessel *Gertrude L. Thebaud* sailed to Washington, D.C., with a crew of New England master mariners to seek federal aid for the New England fisheries. They presented to the President and to Congress a brief entitled, "Statement Setting Forth Serious Situation Confronting Fishermen, Captains, and Vessel Owners at the Port of Gloucester, Mass." The brief pointed out that fishing trips were not even covering operating expenses, exclusive of labor and overhead. (Brief available at Gloucester Fisheries Association, Gloucester, Mass.)

[25] Cf. H. Haberland, "Gloucester — Three Centuries a Fishing Port," *Commercial Fisheries Review* (U.S. Fish and Wildlife Service, Washington, D.C., June, 1946), pp. 1–5.

[26] Redfish, as mentioned above, are known also as "rosefish" and "ocean perch."

[27] Industry representatives are arguing still about claims of discovering the

TABLE 9

Gloucester Landings, 1914–1951

Year	Redfish [a]		Total — all species fresh		All species fresh and salted	
	Millions of pounds	Millions of dollars	Millions of pounds	Millions of dollars	Millions of pounds	Millions of dollars
1914	.. [b]	.. [b]	49.34	1.03	70.25	1.78
1915	.. [b]	.. [b]	49.68	1.01	73.70	1.83
1916	.. [b]	.. [b]	46.52	1.21	66.68	2.16
1917	.. [b]	.. [b]	40.06	1.37	58.14	2.45
1918	.. [b]	.. [b]	62.00	2.20	74.18	3.06
1919	.. [b]	.. [b]	61.62	1.59	71.37	2.15
1920	.. [b]	.. [b]	39.11	1.08	46.74	1.46
1921	.. [b]	.. [b]	26.75	0.66	33.02	0.92
1922	.. [b]	.. [b]	30.40	.54	37.75	.81
1923	.. [b]	.. [b]	29.01	.65	35.03	.91
1924	.. [b]	.. [b]	29.26	.73	35.85	1.04
1925	.. [b]	.. [b]	42.16	1.03	49.47	1.39
1926	.. [b]	.. [b]	49.22	1.23	54.90	1.48
1927	.. [b]	.. [b]	46.06	1.25	52.55	1.49
1928	.. [b]	.. [b]	39.41	1.37	41.90	1.48
1929	.. [b]	.. [b]	49.14	1.52	53.88	1.71
1930	.. [b]	.. [b]	43.66	1.18	47.36	1.35
1931	.. [b]	.. [b]	21.26	0.65	24.85	0.78
1932	.. [b]	.. [b]	23.44	.38	25.33	.43
1933	.. [c]	.. [c]	18.31	.35	21.74	.44
1934	.. [c]	.. [c]	37.30	.68	40.13	.76
1935	2.90	0.03	46.93	.81	51.27	.94
1936	17.09	.24	57.06	1.10	59.41	1.17
1937	17.02	.25	44.70	0.86	46.24	0.92
1938	29.41	.33	60.70	.90	63.01	.97
1939	44.42	.59	75.66	1.20	75.77	1.21
1940	57.40	.85	96.16	1.72	96.18	1.72
1941	99.88	2.01	148.45	3.29	148.45	3.29
1942	91.29	2.74	157.74	6.08	157.74	6.08
1943	83.99	3.40	170.10	9.05	170.10	9.05
1944	91.58	3.56	188.66	9.10	188.66	9.10
1945	102.04	3.96	213.50	11.18	213.50	11.18
1946	130.90	5.84	217.97	10.82	217.97	10.82
1947	95.36	4.20	163.71	7.64	163.71	7.64
1948	176.80	7.41	251.11	11.24	251.11	11.24
1949	169.28	7.30	250.91	10.46	250.91	10.46
1950	120.29	5.69	195.93	9.06	195.93	9.06
1951	177.69	8.83	259.67	12.69	259.67	12.67

Source: U.S. Fish and Wildlife Service. Figures rounded to hundredths.
[a] Or "rosefish" or "ocean perch."
[b] No landings reported.
[c] Negligible landings not reported.

of the new opportunity. It was only a few hours from banks in the Gulf of Maine where redfish were plentiful. It had unoccupied buildings and wharf space, as well as available vessels and fishermen eager to work more regularly. Gloucester was also aided by Boston's limitations. Boston was a beam trawler port, geared for the handling of the larger groundfish, especially haddock and cod. It could not easily accommodate additional millions of pounds of redfish.

Gloucester's fishing industry has expanded phenomenally in the last fifteen years. Redfish landings soared from a few thousand pounds in 1934 to almost 178 million pounds in 1951 and constituted nearly 70 per cent of the total catch.[28] Landings in 1934 were worth but $755,000; in 1951, they brought fishermen and vessel owners $12.7 million. After the second World War the fleet included 248 vessels; in 1933 there were a hundred less in service. Since 1938, the port has had its own fish pier and public freezer.[29] Fishing is not only traditional in Gloucester; it is also the economic mainstay of the city An estimated 70 per cent of the 25,000 population depend, directly or indirectly, upon fishing for their livelihood.

Fish landed at Gloucester have been sold since 1944 through a selling room established and managed by the Atlantic Fishermen's Union. Prior to that year all sales were on an individual bargain basis between the fishing captain and the dealer.

At the outset a committee of Gloucester dealers did work briefly with the union's Gloucester local in setting up rules for the selling room, but the union has managed the exchange unilaterally since its inception. The parties could not produce the rules during a court investigation in 1947; the presumption is that if they were ever reduced to writing they were quickly lost.[30]

---

market possibilities of redfish; but the late John C. Wheeler, for thirty years an executive and a student of the industry, insisted that the first redfish fillets were cut by the General Seafoods Corporation in Boston as the result of a suggestion by Mr. Slade Gorton of Chicago in 1933. The latter was seeking a small fillet to compete with fresh water varieties. Mr. Wheeler writes that J. L. Alphen, General Manager of the General Seafoods concern, orderd the processing of the first redfish fillets.

[28] Data from the U.S. Fish and Wildlife Service.

[29] These facilities were constructed as a joint project by the federal government, the state of Massachusetts, and the city of Gloucester.

[30] Cf. J. Broadhurst, "Findings of Fact and Order for Decrees," *Commonwealth of Mass. v. Patrick J. McHugh et al.*, Equity No. 58644, Suffolk Superior Ct., Boston, Mass., July 31, 1947, page 8.

In the selling room, a union-paid custodian records bids by the dealers for each boatload of fish. The custodian has no authority to effect a sale. Each sale is consummated when a vessel's captain accepts the prices bid by a given dealer for the species included in the captain's trip or boatload. The trips are sold one at a time in chronological order. The custodian shuffles cards containing the vessels' names to determine the sequence of sales. Any captain may refuse all bids for his trip and hold over for a subsequent selling hour. Sales take place daily at 7 A.M. and 9 A.M. Dealers pay no fee for bidding privileges. Up to July, 1947, the union barred from bidding any dealer with whom it had a dispute over the value of the catch purchased at the auction; since that date the Massachusetts courts have ordered the union to allow all buyers to bid on equal terms.[31] The events leading up to this decision and the issues springing from selling room operations will be discussed in Chapter V.[32]

The union set up the selling room system because the fishermen felt that selling to the dealers individually did not bring the full pressure of competition to bear upon prices. The method has never received the wholehearted support of the dealers, but practically all of them admit the need for some sort of fish exchange. They simply do not like the idea of the union's running the auction. For the most part the dealers do not bargain with the union, for the majority of the port's vessels are nondealer owned.

About twenty dealers were doing business in Gloucester in 1949. Approximately one half this number started their operations during the second World War. Although only four or five were of any great size, the port's oldest concern was fully integrated. It handled about one-fifth of the fish landed in the port; it operated vessels and sold salt, smoked, pickled, and canned fish, as well as frozen fish products; it marketed several by-products. The Fishermen's Union has continued to suspect this firm of exercising price leadership in the selling room but only the fact of the firm's size tends to stand in its favor of this charge. No actual evidence to support the suspicion has ever been adduced.

Few data are available to indicate profitability in the Gloucester

---

[31] Cf. *Commonwealth of Massachusetts v. Patrick J. McHugh et al.*, Equity No. 58644, Suffolk Superior Court, Boston, Mass., July 31, 1947.

[32] See section on Union Policy and Public Policy.

industry; only the largest firm publishes statements. Moreover, only the larger firms employ systematic cost accounting methods and maintain adequate records for such purposes. Inquiries by the author in 1948 brought the reply that most of the smaller dealers run their business "out of their hats." [33] It should be noted that Boston suffered from such conditions in the twenties, but benefited greatly from the herculean efforts of their trade association manager to educate the dealers in the profitability of using cost accounting principles.[34]

Several measures do point, however, to very favorable earnings during the second World War period. The Federal Trade Commission found that New England medium otter trawlers increased their annual earnings from $1,881 in 1941 to $13,402 in 1943.[35] This class of vessel has been important in Gloucester; [36] Gloucester owners may not have been as successful as the "typical" New England owner cited by the Commission, but their earnings did encourage expansion of the Gloucester fleet. Between January, 1940 and August, 1944, seventy-seven vessels joined the fleet; only forty-three dropped out. Moreover, the additions greatly expanded the fleet's fishing capacity, for they were larger vessels on the average than those withdrawn.[37] The rise in the average price of fish landed from 2.2 cents per pound in 1941 to 4.96 cents per pound in 1946 inspired an increase of 47 per cent in the volume of annual landings between the two years.[38]

The available evidence indicates that the dealers, too, had exceptional wartime earnings. Five of the firms which operated from early 1942 through 1946 added $238,916 as a group to their surplus ac-

---

[33] Mr. L. J. Hart, Secretary of the Gloucester Fisheries Association, agreed that this was one of the "curses" of the Gloucester industry. Dealers not appreciating fully their real costs often offered prices to buyers which hurt badly those businesses attempting to relate prices quoted to costs of operation.

[34] Cf. E. H. Cooley, "Let's Talk about Fillet Costs," "What is the Cost of Cutting Fillets," and "How to Compute Fillet Costs," *The Fishing Gazette*, April–June, 1931.

[35] Figures are average annual earnings for six typical medium draggers in 1941 and eleven such vessels in 1943. See Federal Trade Commission, *Report on Distribution Methods and Costs*, Part VIII, June, 1945, pp. 18 and 19.

[36] Cf. H. M. Bearse and W. F. Royce, "Fishing Vessels of New England and New York City, 1945," Fishery Leaflet 167, U.S. Fish and Wildlife Service, Washington, D.C., January, 1946.

[37] Cf. "Changes in the New England Fishing Fleet, 1940–1944," *Atlantic Fisherman*, August, 1944.

[38] Data from U.S. Fish and Wildlife Service.

counts, increasing them by 121 per cent in the period.[39] The largest concern, taken separately, reported net profits which averaged $182,440 for 1942 through 1946 after taxes.[40] These earnings represented a gain of about 15 per cent on the average over 1941 earnings.

Between early 1947 and 1949, Gloucester's fishing industry turned far less profitable. The largest dealer's net income slumped from $227,000 annually to $3,403. At least two firms which were opened during the second World War went out of business in 1949; three others were bought out at mortgage value or taken over by existing firms. The vessel owners seem to have had an equally poor experience. Their association reported to a Congressional committee that three out of ten "typical" large Gloucester draggers lost money in 1948; the association's figures indicated that a drop of six per cent in the 4.25 cents per pound average price received for redfish would have wiped out the profits of all ten draggers. Smaller vessels were even closer to the margin, according to the association.[41]

Between 1949 and 1951 things brightened considerably. The data shown in Table 8, while quite fragmentary, show that processors' earnings increased. In late 1952, the United States Tariff Commission reported that records of twelve large and medium trawlers operating from Gloucester earned a net income of only $53,000 on total sales of $1,184,000 in 1949; their net income in 1951 was about $120,000 on sales of $1,330,000. This represented ratios of net income to sales of 4 and 9 per cent respectively in the two years.[42]

[39] Companies include Cape Ann Fisheries, Gloucester Seafoods, Mariners Fish, Producers Fish, and Progressive Fish Wharf. Data from Certificates of Condition filed under the Business Corporation law of Massachusetts, State House, Boston, Massachusetts.

[40] Taxes in some years exceeded final net earnings. See Annual Reports, Gorton-Pew Fisheries Co., Ltd., Gloucester, Massachusetts.

[41] Cf. Gloucester Fishing Vessel Owners' Association, "Statement in Reference to Serious Impact of Vast Quantities of Groundfish Fillets on the Fishing Vessel Owners of Gloucester, Mass.," Subcommittee on Fisheries, U.S. House of Representatives, February 15, 1949.

[42] Data before federal taxes and after managerial salaries. Please note that the net earnings of medium trawlers were far less favorable than that of the group of twelve vessels. The ratio of net income to total sales of medium trawlers was less than zero in 1949 and 1950. In 1951 it rose to 3 per cent. See the Tariff Commission's "Summary of Information Obtained During the Investigation on Groundfish Fillets," 1952, p. 177.

New Bedford

After the decline of the whaling industry, New Bedford did not take its place again as one of the leading seafood marketing centers of New England until after 1937, although a fairly large fishing fleet had used it as a home port since at least 1920. The fleet had formerly landed its catch at New York city.[43]

TABLE 10

New Bedford Landings, 1938–1951

Year	Yellowtail		Sea Scallops		All species	
	Millions of pounds	Millions of dollars	Millions of pounds	Millions of dollars	Millions of pounds	Millions of dollars
1938	6.07	..	4.26	..	21.57	No Data
1939	10.96	..	4.43	..	27.55	1.35
1940	17.52	..	4.41	..	37.40	1.59
1941	28.33	..	5.58	..	46.06	2.34
1942	36.72	1.65	5.45	1.73	57.88	4.55
1943	25.48	1.84	3.83	1.62	62.16	5.98
1944	14.35	0.91	4.01	1.30	74.94	6.28
1945	15.84	.91	3.90	1.28	101.36	8.60
1946	17.13	1.31	8.94	4.10	90.32	12.24
1947	20.82	1.86	10.68	5.28	73.12	10.42
1948	25.21	2.41	10.08	5.28	77.57	11.77
1949	19.65	1.94	11.71	4.30	105.69[a]	9.67[a]
1950	14.06	1.58	11.98	5.59	116.91	11.34
1951	12.62	1.70	12.60	5.64	79.32	11.92

Source: U.S. Fish and Wildlife Service. Data for 1938–1941 inclusive are estimated.

[a] Includes about forty million pounds of red hake landed as "trash fish" to sell for by-products purposes at just under one cent per pound.

New Bedford's phenomenal growth as a primary fish market depended upon a combination of related developments. The introduction of refrigerated trucking service into New York's Fulton Market enabled the city to take advantage of its fishing fleet, its ample docking facilities, and its favorable position with respect to the southern New England fishing grounds. Development of flounder filleting in the port in 1938 and the opening of the city's first freezer exclusively for fish freezing and storage improved New Bedford's marketing posi-

[43] Cf. W. R. Royce, "Fish Production at New Bedford," *Commercial Fisheries Review*, January, 1946, pp. 5–8.

tion.[44] The very favorable prices for fish generated by the second World War — a factor which added to the prosperity of all New England ports — helped New Bedford in particular to expand.

Prior to 1943 only five dealers were operating in New Bedford. One dealer bought over 65 per cent of the fish landed and occupied the position of price leader according to the Federal Trade Commission.[45] In that year, all five dealers were prosecuted by the U.S. Government for violating the Sherman Act. Finding them guilty, the court ordered the dealers to cease monopolizing pier space; to stop allocating among themselves all fish landed in the city; to refrain from agreeing upon the prices they would pay for fish and the margins to be charged in the sale of fish.[46]

The federal antitrust action removed obstacles to the entry of new fish dealing firms; but the Office of Price Administration gave the main impetus to expansion. The price agency set dealer margins at two cents per pound, 150 per cent above the prewar level.[47] Moreover, the agency established ceiling prices on fish at the same level as in Boston,[48] which had been the highest price port in New England. The number of dealers more than doubled in New Bedford between mid-1943 and mid-1944; by 1945 about thirty-one dealers and processors were operating in the port.

New Bedford's seafood specialties are flounders, particularly the yellowtail flounder, and deep-sea scallops. Both sell in national markets. In 1948, yellowtail flounders accounted for one third of the weight and one-fifth of the value of New Bedford landings.[49] Better

[44] Cf. "An Encouraging Development," *New Bedford Standard-Times*, November 17, 1937.

[45] Cf. *Report of the Federal Trade Commission on Distribution Methods and Costs*, Part VIII, June 30, 1945, page 9.

[46] Cf. *U.S. v. L.S. Eldridge and Son Co., Inc.*; *New Bedford Fish Co.*; *Jos. Goulart Fish Corp.*; *Seaview Fish Co., Inc.*; and *Acushnet Fish Corp.*, U.S. District Court of Mass., December, 1943.

[47] Cf. testimony of W. D. Eldridge in *Price Ceilings on Fisheries*, Hearings on H. R. 52, Subcommittee on Fisheries, Committee on Merchant Marine and Fisheries, House of Representatives, 78th Congress, 2nd Session, April 4, 7, 12, 13, 14, May 5 and June 26, 1944, page 56.

[48] Cf. *Report of the Federal Trade Commission on Distribution Methods and Costs*, Part VIII, 1945, page 8.

[49] Fractions computed from data supplied by U.S. Fish and Wildlife Service. As a result of the slump in landings of this species, the proportions were much smaller by 1951.

than 60 per cent of the nation's sea scallop supply comes from New Bedford. Scallops constituted 45 per cent of the value of the 1948 catch, even though they accounted for only 13 per cent of the total weight.[50] As Chapter II pointed out, vessels specially equipped with dredges operate in this offshore shell fishery.

Fish landed at New Bedford are sold through a selling room operated by the Atlantic Fishermen's Union, as in Gloucester. The union initiated sales at auction in 1941 in the open on the city dock;[51] after the second World War a room in the wharfinger's building on the pier became available.

The New Bedford auction system is unique. A selling period of one-half hour is held each morning. Dealers may bid by species on all boatloads[52] of fish simultaneously through the half hour. The union delegate records the bids, as they are made, on a blackboard under the vessel name and beside the type of fish affected by the offer. In bidding on any boat, a given dealer may raise or lower the prices previously offered, so long as his offer raises the total value of the vessel's fish. Any species on the vessel not mentioned in his bid is assumed to be desired by the dealer at the price previously offered. Through this process the several dealers compete to secure the catch of the vessel or vessels in which each is interested. A bell stops all bidding at the close of the half-hour; each vessel's catch is sold to the dealer whose value bid is highest for that boatload of fish when the bell sounds, providing the vessel's captain accepts the offer. He may decline and hold over for the next day's sale. Exhibit I illustrates the operation of this auction market.[53]

Vessel owners and dealers generally belong to two distinct economic groups in New Bedford, just as they do in Gloucester. Both groups enjoyed high earnings during the second World War. Though profit

[50] Percentages computed from data supplied by U.S. Fish and Wildlife Service. The percentages for 1951 were 47 and 15 respectively. Up to 1952 depletion had not affected supplies. Moreover, the Fishermen's Union had limited landings to 100 gallons per man per trip. For an excellent discussion see Premetz and Snow, "Status of New England Sea-Scallop Fishery," *Commercial Fisheries Review*, May, 1953.

[51] Information from letter of Patrick McHugh, Secretary-Treasurer, Atlantic Fishermen's Union to F. F. Johnson, Acting Chief, Div. of Fisheries Industries, Fish and Wildlife Service, Washington, D.C., August 9, 1941.

[52] Listing of quantities of each species in the vessel catch is called the "hail."

[53] For further discussion of primary markets see Chapter V.

data are lacking, the one hundred-vessel expansion of the New Bedford fleet between 1941 and 1945 reflected the attractiveness of boat owners' returns.[54] Landings doubled and landing prices nearly tripled.[55] Between early 1942 and the end of 1946, two of three dealers filing certificates of condition with the state of Massachusetts increased their combined surplus accounts from $68,894 to $197,154 even though war-born companies conducted most of the processing and dealing in the boom years of 1944 and 1945.[56]

EXHIBIT I

The Bidding Process in the New Bedford Selling Room [a]

Bidders	W	W	W	E	W	E	W	E	W (buyer)[b]
Hail	*Opening bid*	*Ensuing bids*							*Final bid*
10 YT	8	8¼	8½	8.65	8.70	8.75	8.75	8.8	8.8
0.4 BB	5								5
0.5 COD	4								4
0.3 FLUKE	15								15
0.2 HAKE							3.0		3.50

[a] Example suggested by Raymond J. Buller, Aquatic Biologist, U.S. Fish and Wildlife Service, formerly in charge Market News Office, New Bedford, Mass.

[b] In this instance, dealer W opened the bidding on the vessel M & K and continued to increase his own bid until dealer E offered competition by increasing the price of yellowtails. E's bid of 8.65¢ on the latter species meant that he was willing to pay the amounts previously bid on the other species. W increased the bid on yellowtails again, and E immediately raised the bid to 8.75¢ per pound. Then W increased the bid on hake to 3¢ per pound. E immediately raised the price of yellowtails to 8.8¢. Just as the bell rang denoting the end of the auction period, W raised the bid on hake to 3.5¢. Dealer W was the successful buyer because his was the highest value bid for the trip. By entering the 3.5¢ bid for hake, dealer W assumed the previously offered prices for the other species. (The quantities of fish listed in the table are in thousands of pounds.)

The Federal Trade Commission warned in 1945 that the danger of overbuilt facilities of all kinds loomed if the wartime mushrooming of the fish business in New Bedford should not prove permanent.[57]

[54] The number of vessels increased from 77 to 177, according to the vessel owners' representative, the New Bedford Seafood Producers' Association.

[55] Data from the U.S. Fish and Wildlife Service.

[56] Cf. testimony of W. D. Eldridge in *Price Ceilings on Fisheries*, Hearings on H.R. 52, Subcommittee on Fisheries, Committee on Merchant Marine and Fisheries, U.S. House of Representatives, 78th Congress, 2nd Session, April 4, 7, 12, 13, 14, May 5 and June 26, 1944, pages 56 and 57.

[57] Cf. *Report of the Federal Trade Commission on Distribution Methods and Costs*, Part VIII, 1945, page 55. The same comments may be made about business practices of the smaller dealers as were made for Gloucester (see page

The reckoning came in 1949. A comparatively sharp drop in landings of food fish after 1946 undermined the profitability of a large sector of the New Bedford dragger fleet. Landings of food fish, except scallops, fell from 81 million pounds in 1946 valued at $8 million to 51 million pounds worth $5.0 million in 1949.[58] This represented the poorest showing of New Bedford since 1943. Many vessel owners claimed to be on the verge of bankruptcy. A singular development saved a large number from that fate. New Bedford took the lead in landing "trash fish," principally red hake. More than 42 million pounds were sold for by-products purposes at an average price of one cent a pound; 1949 marked the first year of fishing for lesser-known, nonpopular edible species and inedible varieties specifically to supply fish meal, oil, solubles, fertilizer and animal food manufacturers, on a tremendous scale in New England. The development will be discussed in Chapter VII.

Other segments of the industry felt the postwar pinch. Scalloping boomed from 1946 through 1948; landings averaged 150 per cent annually above the four million pounds landed in 1945; landed value averaged nearly $5 million against $1.3 million in 1945. The setback came in 1949 when landings nearly 20 per cent higher than the 10 million pounds dredged in 1948 brought the fishermen and vessel owners almost 20 per cent less than the $5.3 million they received in the previous year. One public freezer went through bankruptcy and virtually ceased operations. The other was threatened with mortgage foreclosure by the Reconstruction Finance Corporation. The dealers fared only a little better. At least five companies went out of business in 1949 alone.

Few data are available to show the upturn in conditions between 1949 and 1951, but the sharply better picture presented by catch values is somewhat indicative. They expanded about 23 per cent between the years mentioned. In the main, the improved value of landings resulted from the favorable market for sea scallops, but also we should note that other edible fish (except yellowtail) expanded in both quantity and value as the fishermen turned from the lower-

---

31). If anything, the experience of New Bedford was worse. The profitability of the war period drew many into the business who knew little if anything about fish.

[58] All data on landings and prices from the U.S. Fish and Wildlife Service.

priced by-products species to landing larger quantities of scrod haddock, blackback flounder, and cod.

Portland and Rockland

The marketing of redfish rejuvenated the fishing industry of Portland, just as it saved Gloucester from possible extinction as a major fishing port. Landings of redfish expanded from 70,000 pounds in 1935 to about 37 million pounds in 1950, when Portland's total landings reached the highest point in their history. In 1950, redfish made up 58 per cent of the weight and about 60 per cent of the value of Portland's total landings.[59]

The facilities in Portland have not improved to reflect the expansion in production, as they have in Gloucester and New Bedford. With few exceptions, the docks and buildings used by the industry in 1951 were in poor repair. On the whole, fishing vessels operating out of the port were small in size and number.

The primary market in Portland is not organized as it is in the Massachusetts ports. Fish are sold to the dealers by fishing captains on an individual basis. As we have indicated, the Atlantic Fishermen's Union has had an agent in the area since 1946, but has encountered mediocre success in organizing. One obstacle to successful organization has been the fact that small boat operations tend to dominate in the port; smaller vessel fishermen are exceptionally difficult to organize.

In 1951 there were about nine firms in Portland cutting fillets for the wholesale trade.[60] Although most of the firms were small, two of them represented substantial recent investments by Boston companies. A third, which processes a large proportion of the redfish landed in the port, has been connected with a large midwestern fish wholesaler who handles its products.

Few figures are available to indicate the profitability of the Portland

[59] See Table 11 for data and sources. Redfish landings dropped from 1950 to 1951 because during the former year more vessels, including many from other ports, landed redfish at Portland during June and July when the port of Gloucester was shut down almost completely by a labor dispute involving the Seafood Workers' Union.

[60] Information received from D. A. McKown, Agent, U.S. Fish and Wildlife Service, Portland, Maine.

TABLE 11

PORTLAND LANDINGS, 1916–1951

Year	Redfish		All species	
	Millions of pounds	Millions of dollars	Millions of pounds	Millions of dollars
1916	.. [a]	.. [a]	20.55	0.51
1917	.. [a]	.. [a]	18.57	.74
1918	.. [a]	.. [a]	21.80	.88
1919	.. [a]	.. [a]	21.71	.69
1920	.. [a]	.. [a]	12.75	.62
1921	.. [a]	.. [a]	13.24	.60
1922	.. [a]	.. [a]	15.76	.63
1923	.. [a]	.. [a]	15.22	.69
1924	.. [a]	.. [a]	15.93	.54
1925	.. [a]	.. [a]	18.13	.61
1926	.. [a]	.. [a]	15.96	.57
1927	.. [a]	.. [a]	16.23	.53
1928	.. [a]	.. [a]	17.54	.56
1929	.. [a]	.. [a]	17.44	.60
1930	.. [a]	.. [a]	18.17	.57
1931	.. [a]	.. [a]	18.83	.57
1932	.. [a]	.. [a]	11.29	.28
1933	.. [a]	.. [a]	12.71	.31
1934	.. [a]	.. [a]	16.06	.40
1935	.. [a]	.. [a]	14.48	.34
1936	.. [a]	.. [a]	16.12	.38
1937	.. [a]	.. [a]	17.12	.40
1938	1.16	0.01	18.86	.36
1939	3.87	.04	17.70	.34
1940	6.66	.08	23.43	.47
1941	14.69	0.250	25.68	0.54
1942	13.91	.37	20.54	.68
1943	12.25	.45	18.29	.87
1944	10.47	.38	17.05	.73
1945	11.23	.42	21.96	1.04
1946	20.89	.82	35.61	1.62
1947	13 [b]	.42 [b]	27.43	1.30
1948	20 [b]	.68 [b]	37.56	1.59
1949	24 [b]	.87 [b]	48.93	1.66 [c]
1950	37.18 [b]	1.45 [b]	62.88	2.41 [c]
1951	30.64 [b]	1.40 [b]	56.94	2.36 [c]

Source: U.S. Fish and Wildlife Service. Data for 1949 through 1951 are estimated. Figures are rounded to nearest hundredth. Data generally exclude salt fish as negligible.

[a] No redfish landings reported, 1916–1934. Landings negligible 1935–1937.

[b] Data for Cumberland County. Portland is the largest landing point in that county.

[c] Estimated from data for Cumberland County.

fishing industry; there is some evidence that Maine's future redfish operations may be larger in Rockland than in Portland. In 1949, one of the larger Portland companies opened a modern icing plant in Rockland. The General Seafoods Division of General Foods Corporation, with large plants in Boston and Gloucester, acquired extensive facilities in Rockland in 1946. In that year the Division's chief executive announced plans eventually to handle one hundred million pounds of fish annually at the port. The steady rise in redfish landings shown in Table 12 indicates that the firm is moving in that direction, for

TABLE 12

ROCKLAND LANDINGS, 1947–1951

Year	Redfish		All species [a]	
	Millions of pounds	Millions of dollars	Millions of pounds	Millions of dollars
1947	24.02	.72	29.15	.945
1948	25.36	.91	35.09	1.39
1949	30.39	1.11	39.86	1.48
1950	38.95	1.55	50.79	1.98
1951	41.82	1.96	52.12	2.35

Source: U.S. Fish and Wildlife Service. Data for Knox County, Maine. Rockland is the principal port in that county.

[a] Excluding sea herring, which are not covered by this study. (Sea herring are the basis of the sardine canning industry.)

General Seafoods buys most of the redfish landed at Rockland.[61] In 1949, for example, General Seafoods purchased about twenty-eight million of the approximately forty million pounds of groundfish landed (mostly redfish) in Rockland. One other firm bought most of the remainder.[62]

The Seafoods expansion at Rockland is the second attempt to make the port a major fishing center. The East Coast Fisheries Company, a ten-million-dollar concern promoted by New York interests in 1919, envisaged landings of three hundred million pounds annually

[61] General Seafoods Division was until about 1947 the General Seafoods Corporation, owned by the parent company, General Foods. See the address of J. L. Alphen, President of General Seafoods Corp. reported in "Double Launching of Bluepoint Boats," *Atlantic Fisherman*, August 1946, page 18. Somewhat recently the name was changed again to Birdseye Division of the parent company.

[62] Statement interpolated by the author from information furnished by L. R. Cates, Statistician, Maine Sea and Shore Fisheries, January 28, 1950.

at Rockland.[63] The venture collapsed in 1920 when it could not market quickly and profitably the huge production of its trawlers. The Seafoods expansion, however, is much better grounded. During 1952 that company operated four large trawlers, formerly berthed at Boston, in the redfishery from Rockland.

The Atlantic Fishermen's Union met with about the same success in organizing at Rockland among the port's fishermen as it did in Portland. By late 1952, however, the union reported it had more than one hundred men signed up in the former port, though it had achieved no contracts. By this date, organization had not proceeded far enough for the union to play the role in Rockland that it did in the Massachusetts centers. But further pronounced expansion is bound to encounter increased organizational pressures from the union; indeed, such a development would probably entail drawing at least some experienced union fishermen from the Massachusetts ports. Consequently, the future of Portland and Rockland is nearly as bound up with union policies as that of Boston, Gloucester, and New Bedford.

[63] Cf. "Efficiency Engineers Enter East Coast Enterprises," *Fishing Gazette*, N.Y., October, 1920, pp. 6–8. The entire deal was denounced in the industry as a stock speculation scheme and later events seemed to justify the denunciation. To see how the proposal raised the town's hopes, see the *Rockland Courier Gazette*, August 26 and 29, 1919.

CHAPTER IV

ORGANIZING THE INDUSTRY

ORGANIZATION among the several economic groups in the main New England ports has progressed at greatly varying rates. The experience is summarized in Table 13. Generally, employer opposition counted heavily in delaying union development until after 1937. In turn, the successful spawn and evolution to strength of the unions after that date stimulated greater cohesion among employers. In Boston, they adapted earlier organized trade associations for the new task of collective bargaining; in Gloucester and New Bedford, they initiated group action; in Portland, where union pressures were negligible,[1] they remained unorganized.

The Atlantic Fishermen's Union presents the strongest "united front" in the industry; it is the focal point of our study. This chapter will unfold the details of the fishermen's struggle for organization and the salient features of the union they have built. To complete the spectrum, we will review the organizational experiences of the other principal economic groups.

THE ATLANTIC FISHERMEN'S UNION

The successful launching of this union at Boston in 1937 marked the fishermen's second major attempt in twenty two years to secure continuous, effective representation. Their first organizational effort, the Fishermen's Union of the Atlantic, started in 1915 by William H. Brown, mushroomed during the first World War.[2] At its peak in 1920, that union had represented nearly all fishermen sailing on the larger

---

[1] In 1952 the Seafood Workers' Union claimed to have six firms under contract, however.

[2] Cf. *Constitution and By-laws*, Fishermen's Union of the Atlantic, affiliated with the International Seamen's Union, A.F. of L. (I. H. Feinberg and Son, Printers, Boston, Mass.) A copy is held in the writer's files.

TABLE 13

Organization of the New England Fishing Industry in the Principal Ports

Port	Fishermen	Vessel Owners	Dealers	Seafood Workers
Boston	Atlantic Fishermen's Union SIU–AFL [a] Headquarters Organized 1937	Federated Fishing Boats of New England and New York., Inc. Organized 1931	Massachusetts Fisheries Association Organized 1927	Seafood Workers Union, ILA [b] Boston Local Organized 1937
Gloucester	Atlantic Fishermen's Union Gloucester Local Organized 1939	Gloucester Fishing Vessel Owners Association Organized 1943	Gloucester Fisheries Association Organized 1943	Seafood Workers Union, ILA [b] Gloucester Local Organized 1938
New Bedford	Atlantic Fishermen's Union New Bedford Local Organized 1939	New Bedford Seafood Producers' Association Organized 1938	New Bedford Fillet Dealers Association	Seafood Workers Union, ILA [b] New Bedford Local Organized about 1944
Portland and Rockland	Unorganized (Atlantic Fishermen's Union seeking organization since 1946) [c]	Unorganized	Unorganized	Seafood Workers, ILA [b] Portland Local

[a] Seafarers' International Union.

[b] International Longshoremen's Association, AFL. The Gloucester local has changed its affiliation to the Meat Cutters, AFL.

[c] In Rockland, the Fishermen's Union has organized shoreworkers at the Birdseye (General Seafoods) firm. See Postscript for fishermen's status.

vessels from Gloucester and Boston; it had won its best wage settlement in an arbitration proceeding with the large trawling companies — $130 per man per month plus four dollars bonus on every thousand pounds of fish caught in the winter season and three dollars on each thousand pounds landed during the season from April 1 to October 1.[3] Up to 1917 the fishermen had received but $35 per man per month and five dollars bonus for each one thousand dollars brought by the trip.[4]

Relentless industry opposition and a disastrous strike nearly crippled the union by the end of 1921. Following the 1920 wage award, the organization lost the closed shop controls it had built up through use of the union label; thirty Boston wholesalers secured an injunction forbidding the union from "in any way compelling . . . the plaintiffs . . . to place their business upon a closed shop basis."[5] The costly strike occured in 1921. The fishermen walked out on March first in resistance to a wage cut; they returned to work six months later on terms substantially less favorable than those proposed by the owners in February.[6]

Subsequent attempts to bolster the union failed. In 1924 the officers undertook to establish a producer coöperative. The plan was ambitious. At least two-thirds of the union members were to sign a coöperative marketing agreement; they were to deliver all fish to the union's "Eastern Fish Department," a selling agency which would pool the members' contributions by kind of fish and quality; the new department was to market the pooled catches on the most favorable terms; the proceeds of each pool were to "be divided between the members of such Union, after deducting necessary marketing and handling expenses, in proportion to the contribution of each fisherman or group of fishermen sailing on any vessel contributing to such

---

[3] Cf. "Fishermen's Wage Award," *Boston Evening Transcript*, February 20, 1920.

[4] Cf. "The New Atlantic Fishermen's Union," *The American Fisherman*, Boston, Mass., May, 1939, page 2.

[5] Cf. Interlocutory Decree, *Boston Fish Company et al. v. Fred M. Harty et al.*, Commonwealth of Mass., Supreme Judicial Court, January 30, 1920. Mr. Harty represented the Fish Handlers' Union which coöperated with the Fishermen's Union so that both might secure the closed shop. See also *N.Y. Evening Sun*, March 6, 1920, and *N.Y. Times*, March 10, 1920. The legal actions were called "the Massachusetts Injunction Cases."

[6] Cf. "The New Atlantic Fishermen's Union," page 2.

pool."[7] The plan was to operate on a trial basis from December 1, 1924 through 1929. As a proposal it received widespread publicity,[8] but dealer opposition and the refusal of fishing captains to discontinue selling their own fish short-circuited the scheme. It never achieved the operating stage.[9]

Dormant for the next several years, the dying fishermen's organization led desperation strikes to improve wages and working conditions at Boston in 1933 and 1934. When both failed, the union ceased to exist.[10]

After unsuccessful attempts to reorganize in 1935 and 1936, the Boston fishermen invited the National Maritime Union, CIO, to sign them up in the spring of 1937. By July, the organization was completed; the men held their first meeting, adopted a constitution similar to that of the predecessor union, and elected Patrick McHugh and Austin Powers to lead them. The former, until 1937 a captain in the Boston fleet, became secretary-treasurer — the union's chief executive. Powers, formerly a mate on the Boston boats, became assistant secretary-treasurer and McHugh's subordinate.

The new Atlantic Fishermen's Union began bargaining with Boston's dealer vessel owners immediately; by late 1939 it completed organization of Gloucester and New Bedford fishermen. Its membership jumped from 800 in July, 1937 to 1800 in January, 1938, and reached 2800 in August, 1940.[11] The latest count in the

[7] Excerpt from "Coöperative Marketing Agreement," Fishermen's Union of the Atlantic, 288 State St., Boston, Mass., 1924, p. 5; "marketing and handling expenses" included payment for the respective vessel's charges and operating expenses.

[8] Cf. "A Labor Union Takes a Hand in Sales Problem of an Industry," *Printers Ink*, January 24, 1924; "Minute Men of New England Fishing Industry Seek Way to Prosperity by Coöperation," *Boston American*, November 5, 1924; "Co-operative Selling of Fish Expected to Boom Gloucester," *Christian Science Monitor*, December 15, 1923.

[9] Information obtained in interview with the late William H. Brown, February 2, 1948. This coöperative movement should not be confused with the arrangement, about 1927, between Gloucester mackerel captains and the John Nagle company, a Boston concern, to market all their catch. The latter scheme actually functioned. See "Gloucester Coöperative Plan Works Well," the *Fishing Gazette*, April, 1928, page 23.

[10] Cf. "The New Atlantic Fishermen's Union," *The American Fisherman*, Boston, Massachusetts, May, 1939, page 2.

[11] Figures from report of the Union Audit Bureau to the Atlantic Fishermen's Union, Boston, Massachusetts, April 1, 1939 — August 31, 1940.

spring of 1947 totaled about 4000: about 1700 in Gloucester; about 1000 in Boston; about 800 in New Bedford; and a few hundred in Portland and Rockland, Maine, and New York city.[12]

The union changed its affiliation from the National Maritime Union in December, 1937 to a federal labor union of the American Federation of Labor because the men suspected the CIO union of Communist domination.[13] In September, 1941, the organization accepted a charter from the Seafarers' International Union, AFL, reserving the right to revert to the AFL itself at any time that the International attempted to interfere in the fishermen's affairs.[14] Thus the union is an autonomous body within the Seafarers' International; it has its own headquarters in Boston, branch locals in Gloucester, New Bedford and New York city, and an organizer in Portland and Rockland, Maine.

The headquarters meeting in Boston is the keystone in the hierarchy of union authority; the branches are centers of power for strictly local affairs, but any action they vote must have the approval of headquarters.[15] The constitution specifies weekly meetings at headquarters and in the branches,[16] but the stipulation has been honored more in the breach than in the observance.

The gradual lengthening of periods between meetings at headquarters has increased greatly the responsibility of the secretary-treasurer, for he is empowered not only to execute and coördinate membership policy decisions, but also to exercise all powers granted by the constitution to headquarters.[17] In addition, he handles all collective bargaining matters [18] and represents the union at all official functions. Elected annually by secret ballot of the membership at large, the secretary-treasurer also supervises the similarly chosen

---

[12] Cf. "Findings of Fact and Order for Decrees," *Commonwealth of Mass. vs. Patrick J. McHugh et al.*, Equity No. 58644, Superior Court, Boston, Mass., July 31, 1947, page 5; and testimony of union officials in latter case.

[13] Information from the files of the Atlantic Fishermen's Union.

[14] Cf. *Minutes* of the Meeting of the Atlantic Fishermen's Union, Boston, August 5, 1941.

[15] Cf. *Constitution and By-Laws*, Atlantic Fishermen's Union, Revised, 1944, Article IV.

[16] *Ibid.*, Article X.

[17] *Ibid.*, Article VIII.

[18] Two out of three successive meetings at the port for which he negotiates a contract must approve it, however. See Article XI, *Constitution and By-laws*.

assistant secretary and the branch agents; the latter are elected annually by secret vote in each port.[19] Their principal duties are to supervise branch affairs and finances; as far as possible, they are to make sure that no vessel leaves port unless it has a full union crew.[20]

Captain Patrick McHugh has led the union since its inception and was instrumental in founding the organization. An aggressive leader, he has taken in stride the growth in his responsibilities noted above. Nevertheless, like the fishermen whom in a sense he epitomizes, he is conservative and therefore he has not run the union between meetings more or less to suit his own ideas. This to a good extent helps to explain his success in continuing as the chief executive, for the independently-minded fishermen would not long tolerate one who took too much on his own shoulders. We should note, too, that McHugh as a fighting personality has over the years held the respect and admiration of the men. In the tough battles to secure effective organization, McHugh stood up foursquare to the owners, particularly the Boston dealer-vessel owners. Indeed, he has never flinched in the face of employer pressure, and the men know it.[21] The owners and dealers respect McHugh as a fighter; but they think he is uncompromising and unwilling to let by-gones be by-gones. They deplore what they describe as McHugh's unwillingness to encourage the better men to be active in union affairs.

The union's experience with personalities in lesser union offices has generally been not too satisfactory. It has had difficulty turning up good officer material, a problem we shall deal with in Chapter VIII.[22] For the moment, however, we should note that the present organizational structure with power centralized mainly in the Boston membership meeting, has helped to produce this result. Then, too, the tendency of the men to look to McHugh as the man who can get what they want — and he has produced results — discourages some-

[19] Cf. *Constitution and By-laws*, Article VII. Article VI specifies that candidates for office must be American citizens, full members of the union for at least two years, and in good standing for at least one year prior to their nomination. They must not be members of any "subversive organizations." The latter is not defined, but has meant generally communist organizations.

[20] *Ibid.*, Article VIII.

[21] In a very real sense, McHugh has been regarded by the men as their champion.

[22] Some of the port agents have left office under a cloud — suspected of mishandling finances.

what the emergence of new vigorous leadership talent. It should be stressed, however, that this situation is not one engineered by McHugh to perpetuate a regime. It has grown almost naturally from the history of the conflict combined with the internal governmental arrangements of the union. A situation which harassed the union in 1947 and 1948 shows further the desirability of broadening the union's leadership structure. The assistant-secretary treasurer, disagreeing with some of McHugh's policies, withheld his full coöperation. McHugh, in turn, had lost all respect for the former because McHugh felt that his assistant had been personally responsible for an organizational failure in Maine. The stalemate, which ran for almost two years until the membership turned the assistant out of office in 1948, greatly hampered each officer's efficiency.

The Fishermen's Union admits practical fishermen, nonvessel-owning captains, and in Boston, "lumpers" to membership. The latter are the "fish longshoremen" who assist the fishermen in unloading their catch.[23] A substantial number of the union fishermen are naturalized citizens from Canada, Newfoundland, Italy, Portugal, and the Scandinavian countries, where they were born to the sea. Many, like Captain McHugh, have been fishing since they were about twelve or fourteen years old.[24] Italians and Portuguese predominate in Gloucester; Newfoundlanders, Canadians, and Scandinavians in Boston and New Bedford.[25] How the union has tightened its admission requirements in recent years will be explained in the next chapter.

The fishermen are highly individualistic, very "short-run" in outlook, and in a sense, "crisis-minded." Having spent long hours at sea battling the ocean and elements — and surviving often only through his own resourcefulness — the average fisherman has a powerful sense of self-reliance. Moreover, he is quite content to let tomorrow take care of itself; it is the "here and now" which concerns him. Nevertheless, he is not easy to arouse unless the situation is really "tough." Given what he considers a "crisis," however, he will back

[23] Those lumpers who substitute for the fishermen are paid directly by the men they replace. There have been instances of strife within the union between the lumpers and fishermen over "wages," but it has usually cleared up quickly.

[24] McHugh came to the United States from Newfoundland over thirty years ago.

[25] Information from the Atlantic Fishermen's Union.

his leadership unflinchingly. Despite the long standing rule of the profit-sharing lay in the industry, the fisherman identifies himself unqualifiedly as an employee — a wage-earner differently paid from most workers, but still a wage-earner. With his fellows, he will fight hardest when he is convinced that an individual or group stands in the way of what the fishermen consider "fair" wages.

To a large extent, the ingrained bitterness which featured dealer-fishermen relations in the twenties and thirties — a period when nearly every fisherman was "positive" that the dealers were depriving him of a decent wage — has given the industry a heritage of conflict which colors all buyer-Fishermen's Union dealings. For one thing, it helps make the organization exceptionally cohesive when a contest with the dealers over some issue rises to the fore. Thus at the same time that the men are united against any outside force, they will not discipline individuals within the organization to assure conformity with rules they themselves pass. Since 1947, for example, Boston fishermen have had a self-imposed rule to fine anyone who delays a voyage by failing to appear at sailing time. In spite of many violations, no fines have been levied. From this one can see that one of the most difficult officer tasks is to make the membership live up to contractual obligations.

In summary, the fishermen view their union as a vehicle to protect their "rights" and "wages." They tend to view union dues as a sort of insurance premium — $2.50 per month up to 1952 and $3.50 monthly since that time. They do not as a whole conceive of the union as an institution which might play an important role in positive programming for the industry.

SHOREWORKERS' UNIONS

Nearly all shoreworkers [26] in the major New England fishing ports are organized. The International Longshoremen's Association, AFL, has separate Seafood Workers' locals in Boston, Gloucester, and New Bedford; a fourth local in Portland is supervised by the Boston local. One other union, Local 59 of the Teamsters' Union, AFL,

[26] Shoreworkers include those who unload, process, package, and freeze fish in each port. However, the "lumpers" who unload fish in Boston are members of the Atlantic Fishermen's Union.

exerts a powerful influence in New Bedford. Its members transport virtually all fish leaving New Bedford for the New York market.

None of the seafood workers' organizations is very old. The Boston and Gloucester locals were formed early in 1938;[27] the New Bedford local displaced Local 609 of the Amalgamated Meat Cutters, AFL, in 1944. The latter had started in the port about 1940.[28] The Portland local functioned ineffectively from about 1940 until 1946, when it became attached to Boston.[29] The New Bedford Teamsters' local organized the port's fish transport workers during the New Bedford

TABLE 14

HOURLY WAGE RATES FOR SELECTED JOB CLASSIFICATIONS IN BOSTON FILLETING PLANTS, 1939–1940 AND 1947–1952

| | Job classification | | | | | |
Year	Cutters	Scalers	Wrappers	Weighers	Floormen	General helper
1939	$0.82	$0.67	$0.54	$0.57	. . .	. . .
1940	.82	.67	.54	.57	$0.70	. . .
1947	1.32¼	1.14	.98	.98	1.24	$1.06
1948	1.42½	1.23	1.05	1.05	1.33	1.14
1949 [a]	1.47½	1.27	1.09	1.09	1.38	1.18
1950 [b]	1.50	1.29	1.11	1.11	1.40	1.20
1951 [c]	1.56½	1.34	1.15	1.15	1.46	1.25
1952 [d]	1.61½	1.38	1.19	1.19	1.51	1.29

Source: Compiled from data supplied by U.S. Bureau of Labor Statistics. Compilation by U.S. Tariff Commission.

[a] Effective May 3, 1949.
[b] Effective May 3, 1950.
[c] Effective Jan. 8, 1951.
[d] Effective May 3, 1951.

expansion of the later 1930's.[30] Each of the seafood workers' locals and the teamsters' organization follows its own course on policy matters.

[27] Information from the business agents of the Boston and Gloucester locals.

[28] Cf. "Seafood Workers Offer Contract," the *New Bedford Standard-Times*, May 10, 1944.

[29] Information from John Donegan, President and Business Agent, Boston Seafood Workers'. The N.L.R.B. found the former head of the Portland local guilty of establishing a company-dominated union after his removal from leadership of the Portland Seafood Workers'. See "N.L.R.B. Upholds A.F.L. Charges in Maine Fish Case," *Boston Traveler*, March 11, 1949.

[30] See Chapter III.

The seafood workers' unions include such shoreworkers as fillet cutters, general help, and packers. A large proportion of the latter are women. Straight wage systems prevail in the Massachusetts ports, whereas many of the nonunion workers in Portland and Rockland work under piece-rate systems. Tables 14 and 15 show the wages paid

TABLE 15

INDEX OF WAGE RATES IN BOSTON FILLETING PLANTS, UNADJUSTED AND ADJUSTED FOR PRICE CHANGES AND FOR EARNINGS OF FACTORY WORKERS

Year	Wage rates in filleting plants	Consumer prices	Average hourly earnings of factory workers	Index of wage rates in filleting plants, adjusted in terms of —	
				1940 consumer prices	Hourly earnings of factory workers
1940	100	100.0	100.0	100.0	100.0
1948	184	170.2	211.8	108.1	86.9
1952 (January)	208	189.1	248.3	110.0	83.8

Source: U.S. Tariff Commission.

to these workers in Boston since 1939 and demonstrate also how well those workers have fared in terms of real income over that span of years. Wages in Gloucester have run about 10 per cent below Boston, in New Bedford not quite 15 per cent below Boston, and in Portland about 20 per cent lower than in the Hub. The Boston workers have had further advantages over those in the other ports, however, in that they have enjoyed superior "fringe" benefits and, among other things, have had a forty-hour weekly guarantee for some years, whereas shoreworkers in the other ports have not.

Within the structure of markets through which fish flow, the interests of shoreworkers are somewhat closer to those of the dealers than the fishermen. Their dependence upon the fishermen's ability and willingness to land fish in volume has brought different reactions from the seafood workers' locals and the teamsters. The former have sought, with little success, to promote a Joint Council of the Atlantic made up of officers from the Fishermen's Union and the Seafood Workers' locals in all New England ports. The Council operated briefly in 1946 as an agency for unionizing New England's unorganized fishermen and seafood workers, but it became dormant quickly. The officers of the Fishermen's Union were too preoccupied with their own problems to continue the arrangement. The plain fact is that the fishermen have far more bargaining power than the seafood workers in the structure of the New England fish market. The fishermen are not so easily replaced; they can land fish at different ports in the event of a shoreworkers' tieup; they have a tightly-knit organization controlled from Boston, whereas the seafood workers' locals are on their own in their respective ports.

The Teamsters' Union has employed other tactics in New Bedford. It has agitated for "home rule" — separation of the New Bedford fishermen's local from the Boston-directed Atlantic Fishermen's Union. In early 1944, the Teamsters issued a charter for the Greater New Bedford Independent Fishermen's Association.[31] The new organization was to "break" a New Bedford fishermen's walkout in sympathy with Boston against the ceiling fish prices set by the wartime OPA. The strike ended before the "independent" made much headway, but the incident is a reminder to the Atlantic Fishermen's Union that the New Bedford Teamsters' local may contest any work stoppage which interrupts the employment of its members.

EMPLOYER ASSOCIATIONS

VESSEL OWNERS

Boston, Gloucester, and New Bedford vessel owners deal with the Atlantic Fishermen's Union through their respective associations. Alleged "whipsawing" by the fishermen's organization provided the stimulus for group bargaining by the employers in each case.

In Boston, the Federated Fishing Boats of New England and New York, Inc. is the trawler owners' representative. Originally founded in 1931 as a nonprofit corporation to "promote the interests of its members and the interests of the fishing industry generally," [32] the Federated assumed the bargaining role in negotiating the first contract with the port's fishermen in 1939 after it appeared that the

---

[31] Cf. *New Bedford Standard-Times*, March 4, 1944.

[32] Cf. Federated Fishing Boats of New England and New York, Inc., *By-laws*, Administration Building, Boston Fish Pier; also, H. F. Turner, "N.E. Vessels Unite," *Fishing*, New York, Dunlop Publishing Co., June, 1931.

union would extract the best possible terms from each owner if the latter did not bargain as a group.

Despite its age, the Federated is not very cohesive. Its executive secretary, who is not a boat owner, leads the negotiating committee, but each concern signs its own contract as a Federated member when general agreement with the union appears to have been reached. A member may choose to attempt to settle with the union on terms different from those reached by the union and the Federated.[33]

Voting power within the Federated is roughly proportional to the number of vessels which a company has registered and upon which it pays dues.[34] Up to 1949, when the General Seafoods Division of General Foods Corporation sold half its fleet to the U.S. Army for use in Germany, that company had twice the number of vessels registered by the second largest dealer-vessel owner. Since that time, General Seafoods has shared the leadership with two other concerns which also buy and process fish. Together the three firms controlled 16 of the 35 trawlers making up the Federated membership in April, 1950.[35]

The Federated has had two executive secretaries in its span of years. The first, who served until 1945, clashed so often and violently with the Fishermen's Union head that many people blamed personal differences for the mutual hostility of their respective organizations. While it is true that these men had little love for one another, the fact is both personalities "grew up in the business" and mirrored the sentiments and positions of their respective partisans.[36] Under the trying circumstances which followed with such conditions prevailing, the personal animosities of two capable and eager men simply became an almost natural consequence. When the present executive secretary of the Federated entered the scene in 1945, he brought a fresh outlook, uncolored by the bitter past.[37] Nonetheless, little

[33] See Chapter V for discussion of Boston's collective bargaining experiences.

[34] Cf. Federated Fishing Boats, *By-laws*, Article IV, "Voting" and Article V, "Application Fees and Dues."

[35] Cf. Testimony of T. D. Rice, Executive Secretary, Federated Fishing Boats, in *Northwest Atlantic Fisheries Convention*, U.S. Senate Hearings, 81st Congress, 2nd Session, April 4 and 5, 1950.

[36] The first executive secretary was Mr. E. H. Cooley of Newton, Mass. Cooley stated that he was invited to join the industry and start the "Mass. Fisheries" in December, 1927.

[37] The second executive secretary is Mr. Thomas D. Rice of Belmont, Mass.

change in organizational relationships came about immediately, even though the personal relations between McHugh and the present secretary were good.

The reason for this was partly that there really had been remarkably little change in the personality makeup of the Federated over the preceding fifteen years. The principal executive of one of the largest and most profitable dealer-vessel owners — and therefore a man carrying great weight in the port — was vitriolic in his dislike for the union and everything connected with it.[38] The battle scars of the past were just too deeply imbedded to have healed in so short a time — especially when the market clashes we shall describe in Chapter V were constantly present as an aggravation.

The New Bedford and Gloucester employer associations differ from Boston's principally in the character of their membership. Virtually all subscribing vessel owners in the former ports are not fish dealers. Their collective bargaining experience with the Fishermen's Union has departed sharply from Boston's, as Chapter V will indicate.

In New Bedford, the Seafood Producers' Association[39] started informally in 1938 when the Atlantic Fishermen's Union began organizing the port. The Association negotiated its first agreement with the Fishermen's Union in 1940, but remained largely a "paper" organization until 1945. In that year it reorganized and hired a business agent who raised its membership from 25 to more than 180. The Association became the industry's chief spokesman in the port.

The Seafood Producers' members rely rather heavily upon their business agent; he plays a prominent role both at the bargaining table and in settling grievances with the Fishermen's Union, although he is strictly accountable to the membership. The Association's by-laws provide that the president must at some time have been actively engaged as a fisherman.[40] Moreover, they give a fifteen-man board of

[38] The stormy personality referred to is Mr. Bart Whalen of the R. O'Brien Co.

[39] Information on the Association obtained from Captain Daniel F. Mullins, known as the "father" of the port's fishing industry and the founder and first president of the Association; also from Captain John Murley, past president of the Association and the port's authority on fishing vessel appraising.

[40] Cf. *By-Laws*, Seafood Producers' Association of New Bedford, Massachusetts, Article V.

directors, representing equally the groundfishing, inshore, and scalloping sectors of the port's fleet, the power to manage fully the affairs of the Association.[41]

The Gloucester Fishing Vessel Owners' Association [42] was formed in the spring of 1943 to solidify the bargaining position of the cities' vessel owners and to insure uniformity of treatment from the Fishermen's Union. Previously, the union had negotiated "key" contracts with certain vessels in the port's several fleets [43] on an individual basis. Other operators met the standards set, often without written agreements. That system broke down when the vessel owners began to suspect that the union was using the "old Army game" of playing one operator off against another.[44]

The Association functions informally; it retains a lawyer as secretary to help in bargaining and to protect the general interests of its members, but it maintains no offices. Each vessel owner continues to sign his own agreement with the union, but on the basis of the general terms reached by union and Association representatives and announced at a general meeting to the Association's members.

DEALER ORGANIZATIONS

Among the main port dealers, the Massachusetts Fisheries Association in Boston is the only formal association for collective bargaining. Formed in 1927 to promote Boston's fish trade, the "Mass. Fisheries" has negotiated with the Seafood Workers' Union for most of the dealers on the Boston fish pier since 1939.[45] The

[41] *Ibid.* Article III.

[42] All information from M. I. Bernstein, Secretary of the Association and from his files.

[43] The fleets run on nationality and product lines, e.g. the Italian fleet, the Portuguese fleet, the mackerel fleet, the whiting fleet, etc. The whiting fleet has its own business manager.

[44] The expression is that of Mr. M. I. Bernstein, Secretary, Gloucester Fishing Vessel Owners' Association.

[45] The chain stores (A & P, First National Stores, Inc., etc.) negotiate their own agreements individually. They do not belong to the Massachusetts Fisheries, principally to avoid any antitrust law violation which membership might involve. The General Seafoods Division of General Foods Corp. is a leader in the Massachusetts Fisheries. However, General Seafoods has negotiated separately with the Seafood Workers' since that union successfully reorganized the company's workers in 1948. The Seafood Workers' had been bargaining agent briefly before 1943.

organization parallels the Federated Fishing Boats of New England and N.Y., Inc. which we have already described. Both have similar by-laws and the same executive secretary. Since the large vessel owners in the port are also the most prominent dealers, they lead in both the Massachusetts Fisheries and the Federated.

The Gloucester situation is unique. Nearly all dealers belong to the Gloucester Fisheries Association. The latter advertises the port's products but is forbidden by its articles of association to engage in collective bargaining.[46] Nevertheless, the member dealers, with one or two minor exceptions, employ jointly the Association's lawyer to assist a committee they establish to bargain with the Seafood Workers' Union.

In New Bedford the informality of dealer organization reaches the extreme. The New Bedford Fillet Dealers' Association remains inactive, except for periodic negotiations with the Seafood Workers' local. Each firm reaches its own final agreement with the union.[47]

[46] Cf. Gloucester Fisheries Association, *Articles of Association*, Article XI "Prohibition Against Collective Bargaining." (Document Available at 120 Main St., Gloucester, Mass.)

[47] Information from Joseph Sylvia, Business Agent, New Bedford Seafood Workers' local.

CHAPTER V

PROFIT-SHARING AND CONFLICT

THE PROFIT-SHARING LAY SYSTEM under which New England's main port fishermen earn their living has had a stormy experience since the formation of the Atlantic Fishermen's Union in 1937. The essential struggle associated with the lay has been between the fishermen as sellers and the fish dealers as buyers of the catch.

In Boston, the integration of fish dealing and trawler owning, pointed out in Chapter III,[1] has brought the conflict into the collective bargaining process; the groups have clashed both as employers and employees and as buyers and sellers. In Gloucester and New Bedford, where most of the dealers do not own boats,[2] the union has fought with the dealers for higher catch values in the primary market. It has set up the selling rooms, described in Chapter III,[3] to foster greater competition among dealers for fish. In all three ports, it has developed manpower controls and production regulations designed to keep fishermen's incomes at what the men consider a fair level. The state of Massachusetts has intervened through antitrust action to modify some of these union policies.

Why is the lay system used in the industry? What are its peculiar features? How has it figured in the performance of collective bargaining in the respective ports? What specific forms have Fishermen's Union policies taken? How are they related to the buyer-seller controversies between the dealers and the fishermen? In what way have the policies been regulated by Massachusetts public policy?

The following pages will undertake to answer these questions. Those interested in a more formal economic analysis should consult the Appendix.

[1] See page 25.
[2] See Chapters III and IV.
[3] See pp. 29–30 and 35–36.

The "Profit-Sharing" Lay

The deep-sea trawler fishermen's occupation is arduous, isolating, and risky. Their work calls for spending most of their waking hours away at sea; it entails long hours braving foul weather; it requires high physical stamina and strength; it exposes them to accident and death. The risk of life and limb has decreased since the day of the sail-powered vessel, but it still remains high in the modern diesel-trawler fishery.[4] The use of heavy power equipment and machinery on modern trawlers requires alert and nimble coöperation at close quarters if accidents are to be avoided and the trip is to be a success. Furthermore, the fishermen must seek out their elusive, perishable prey; they must take pains in preserving it. The latter factor is especially crucial in fresh and frozen product marketing.

These volatile aspects of the work have led the industry, as far back as colonial days, to use a highly geared incentive system, called the lay.[5] Moreover, some form of share plan rather than straight wages seems to be characteristic of most fisheries operations. An NRA study of fishermen's wages in the early 1930's pointed out the great diversity of share methods whereby fishing vessel crews are compensated, but it concluded that crew members in the marine fisheries are most commonly paid a direct share in the value of the catch.[6]

The current variant of the lay system has prevailed in the New England vessel fisheries since 1928.[7] Chapter IV has described briefly the form employed from the turn of the century up to that year.[8] It gave the fishermen a small monthly wage plus a share in the catch. Neither employers nor fishermen have made a serious attempt to reinstate that method, although during the second World War in Boston in order to achieve some gains within the framework of the

[4] Cf. Robert Moore, "Growing Cost of Protection and Indemnity Insurance," *Atlantic Fisherman* (Goffstown, N.H.), April, 1941, pp. 5 and 23. Safety at sea is a continuing problem of the Fishermen's Union as later pages in this chapter will reveal.

[5] Cf. Raymond McFarland, *History of the New England Fisheries*, page 162.

[6] Cf. J. R. Arnold, *The Earnings of Fishermen and Fishing Craft*, NRA Industries Studies Section, Division of Review, Washington, D.C., 1936, page 54.

[7] Cf. "The New Atlantic Fishermen's Union," *American Fisherman*, Boston, Massachusetts, May, 1939, page 2.

[8] See Chapter IV, page 44.

federal wage-price control program, the union did attempt unsuccessfully to secure what would have amounted to a straight piece-rate system.[9]

Under the present lay arrangement, certain trip expenses are deducted from the gross value of the vessel's catch, called the "gross stock." These charges are "above-the-line expenses." The remainder, called the "net stock," is divided between the vessel owner and the crew in stipulated percentages. The principal running expenses of the trip, called "below-the-line expenses," are deducted from the crew's gross share and the remainder is split evenly among all members of the crew, including the captain. The latter receives, in addition, a percentage of the owner's share. The specialists aboard, such as chief engineer, mate, and cook receive a bonus in addition to a full share in the catch. These payments are known as "pers" in the industry and are usually deducted from the gross stock. The actual lays prevailing are exhibited later in this chapter.[10]

Six variables govern the net share received by each crew member under the lay system: (1) the quantity of the catch; (2) the unit price received for it; (3) the size of operating expenses; (4) the proportion of the "above-the-line" and "below-the-line" expenses; (5) the ratio of the fishermen's share to the "net stock"; (6) the number of men in the crew. In a sense, the last three factors constitute the "shell" of the lay. Set in advance, they provide the framework within which later determination of items (1) to (3), unknown before the start of any given trip, will yield the fisherman's actual share trip by trip. Thus the lay distinguishes the fisherman from nearly all other wage earners. Unlike the barber and the taxicab driver, who also work on shares, the fisherman does not know in advance what price his product will bring. Unlike the factory piece-worker, the fisherman must find his "material"; it is not provided in a continuous stream and its abundance cannot be foretold in advance of any voyage.

Do these uncertainties make the fishermen coventurers with the vessel owner, or employees of the latter? The answer is important mainly because it affects the fishermen's legal rights. The courts have held that the fishermen are employees.[11] They do not own the fish

---

[9] See, however, the Postscript to this book.
[10] See pp. 69, 81, 82 and 83.
[11] Cf. J. R. Arnold, *The Earnings of Fishermen*, page 57.

which they catch.[12] They are entitled to the benefits of the Social Security Act [13] and, up to 1947 at least, the protection of the National Labor Relations Act.[14] Also, since 1946, the fishermen have been eligible for unemployment compensation benefits.

Regardless of whether one calls the fishermen "coventurers" or "employees," however, the strategic consequence of the lay system is the piquing of the fishermen's interest in the related product market [15] as well as in production policies, the costs of vessel operations, and the division of the proceeds of each fishing trip. The derivation of the fishermen's incomes from the value of the product is direct and clear. The fishermen, as their union's counsel put it, are "directly interested in the profits of the voyage." [16] Until the advent of strong organization, the fishermen had little voice in any of these matters. The development of the Atlantic Fishermen's Union has brought forthright action, as later sections of this chapter will reveal.

The lay arrangement has other important effects. Every trip is a separate venture so that the employment period for any fisherman on a given vessel may be as short as ten days. This facilitates part-time employment and a high turnover rate. One study of turnover on six Boston vessels for January through May, 1943, disclosed that 211 men served during the period in the 102 available jobs.[17] To some extent this inquiry reflected the turbulent personnel conditions of a war period and, possibly, an upward bias in its findings because the figures were submitted by employers in opposition to coverage for the

[12] Cf. *Bishop v. Shepard*, 23 Pick. 492, 494–495; and *Adams v. Augustine*, 195 Mass. 289, 81 N.E. 192, cited in *Commonwealth v. McHugh*, 93 N.E. 2nd 751.

[13] Cf. *O'Hara Vessels v. Hassett, Collector of Internal Revenue*, U.S. District Court of Mass., April 23, 1942.

[14] Cf. *In the Matter of Trawler Maris Stella, Inc.*, 12 N.L.R.B. 415; *In the Matter of Federated Fishing Boats of New England and New York, Inc.*, 23 N.L.R.B., 218. There have been no cases under the Taft-Hartley law.

[15] For a discussion of trade union interests in related product markets, see J. T. Dunlop, *Wage Determination Under Trade Unions* (New York: Augustus Kelley, Inc., 1950), chapter VI.

[16] Cf. Henry Wise, Esquire, "Brief for the Appellants," *Patrick J. McHugh et al. v. Commonwealth of Massachusetts*, Equity No. 4897, Supreme Judicial Court of Mass., February, 1950, page 102.

[17] Cf. Brief by E. H. Cooley, representing Boston vessel owners, in *Unemployment Compensation for Merchant Seamen*, Committee Prints Nos. 1 and 3, Committee on Merchant Marine and Fisheries, U.S. House of Representatives, 78th Congress, 1st Session, June 24–29, 1943, pp. 38–41.

fishermen under unemployment compensation laws. But a U.S. Bureau of Labor Statistics study [18] of the Boston fleet for the peacetime year of 1948 yielded similar results. The Bureau reported that 46 per cent

TABLE 16

DISTRIBUTION OF FISHERMEN IN BOSTON FISH PIER FLEET BY ANNUAL EARNINGS AND OCCUPATION, 1948

Annual earnings (dollars)	All fishermen	Captains	Chief engineers	Second engineers	Mates	Cooks	Deck hands
Under 500	212	..	4	12	4	12	180
500–999	120	1	2	5	3	10	99
1,000–1,499	87	3	1	8	3	3	69
1,500–1,999	72	2	3	1	2	7	57
2,000–2,499	81	1	..	4	3	6	67
2,500–2,999	78	..	..	7	1	6	64
3,000–3,499	90	4	2	6	4	4	70
3,500–3,999	73	2	..	3	..	5	63
4,000–4,499	78	..	5	6	3	6	58
4,500–4,999	92	3	2	5	5	5	72
5,000–5,499	93	..	8	2	6	3	74
5,500–5,999	83	3	2	6	7	5	60
6,000–6,499	52	2	5	5	7	3	30
6,500–6,999	46	2	5	3	2	3	31
7,000–7,499	42	1	6	4	1	4	26
7,500–7,999	26	1	1	3	5	2	14
8,000–8,499	23	1	2	..	2	1	17
8,500–8,999	17	2	2	3	3	1	6
9,000–9,499	8	1	3	1	..	1	2
9,500–9,999	5	3	..	..	..	..	2
10,000–10,999	5	3	..	..	2	..	..
11,000–11,999	6	5	..	..	1	..	..
12,000–12,999	5	5	..	..	..	..	..
13,000–13,999	3	3	..	..	..	..	..
14,000–14,999	6	6	..	..	..	..	..
15,000 and over	13	13	..	..	..	..	..
Number of workers	1,416	67	53	84	64	87	1,061
Average annual earnings	$3,676	$9,957	$5,239	$3,692	$4,991	$3,342	$3,149

Source: "Income of Fishermen — Boston Fish Pier Fleet — 1948," U.S. Dept. of Labor, Bureau of Labor Statistics, Washington, D.C., 1949.

[18] Cf. Kermit Mohn, "Incomes of Fishermen — Boston Fish Pier Fleet — 1948," U.S. Dept. of Labor, Bureau of Labor Statistics, Washington, D.C., 1949. See also "Annual Earnings of Boston Fishermen in 1951," U.S. Department of Labor, Bureau of Labor Statistics, Washington, July, 1952.

of the 1416 Boston fishermen worked less than 151 days and only about 25 per cent worked more than 240 days during that year. Nearly 90 per cent of the port's fleet was in operation 240 days or more.

TABLE 17

DISTRIBUTION OF FISHERMEN IN BOSTON FISH PIER FLEET BY ANNUAL EARNINGS AND OCCUPATION, 1951

Annual earnings (dollars)	All fishermen	Captains	Chief engineers	Second engineers	Mates	Cooks	Deck hands
Under 500	85	1	2	11	. .	8	63
500–999	40	1	. .	2	2	4	31
1,000–1,499	28	. .	. .	2	1	5	20
1,500–1,999	42	. .	. .	4	. .	2	36
2,000–2,499	41	. .	2	3	1	3	32
2,500–2,999	34	. .	1	5	. .	2	26
3,000–3,499	43	. .	1	1	3	3	35
3,500–3,999	55	. .	1	5	2	4	43
4,000–4,499	84	. .	. .	5	3	6	70
4,500–4,999	105	. .	4	7	6	4	84
5,000–5,499	133	. .	11	10	7	12	93
5,500–5,999	109	1	5	5	6	7	85
6,000–6,499	102	1	6	7	5	7	76
6,500–6,999	76	1	8	5	4	4	54
7,000–7,499	36	. .	4	3	4	5	20
7,500–7,999	17	. .	3	5	4	. .	5
8,000–8,499	7	2	1	. .	3	. .	1
8,500–8,999	5	2	1	1	1	. .	. .
9,000–9,499	3	1	. .	. .	2	. .	. .
9,500–9,999	8	2	. .	. .	6	. .	. .
10,000–10,999	5	5	. .	. .	. .	. .	. .
11,000–11,999	9	8	. .	. .	1	. .	. .
12,000–12,999	5	5	. .	. .	. .	. .	. .
13,000–13,999	3	3	. .	. .	. .	. .	. .
14,000–14,999	5	5	. .	. .	. .	. .	. .
15,000 and over	12	12	. .	. .	. .	. .	. .
Number of workers	1,092	50	50	81	61	76	774
Average annual earnings	$4,720	$12,063	$5,621	$4,177	$6,138	$4,008	$4,202

Source: "Annual Earnings of Boston Fishermen in 1951," U.S. Department of Labor, Bureau of Labor Statistics, Washington, D.C., July, 1952.

The previously mentioned Bureau of Labor Statistics study, together with a similar investigation conducted in 1951, represented the most detailed survey of fishermen's earnings ever made. The earnings reported are broken down by income brackets and occupation in Tables 16 and 17. Generally, the aforementioned irregularity of employment greatly influenced the Boston fishermen's annual incomes. In 1948, deckhands, or ordinary fishermen, accounted for about three-quarters of the work force; they had average annual earnings of $3149. Yet, not quite one-fourth of this group worked more than 240 days and their average annual pay amounted to $6208. Nearly one-third of the deckhands had 90 days of employment or less; in almost all cases their earnings were under $2000. Average annual earnings of cooks, mates, and second engineers reflected similarly the impact of employment irregularity. Captains and chief engineers, with much more stability of employment as well as superior pay arrangements, showed higher earnings. Captains averaged $9957 and over half their number were paid at least $10,000. Over half the chief engineers worked more than 240 days and averaged $6900 for the year. The part-time employment and limited earnings of fishermen in 1948 seem to have been voluntary to a good extent. Boston owners complained bitterly during that year that their trawlers operated considerably less than full time for lack of crews.[19]

Between 1948 and 1951 average annual earnings advanced sharply. The increase reflected a decline in turnover and an increase in average number of days worked for every occupational group.[20] Table 18 compares the two periods. In every occupation average days worked in 1951 exceeded the experience of 1948. Also, in 1951 nearly two-thirds of all fishermen worked 180 days or more, whereas in 1948 about 63 per cent worked less than 180 days. Consequently, average annual earnings reached $4720 in 1951; they were only $3676 in 1948. In the deckhands group, part-time employment fell the most and earnings rose the most between the two years. Deckhand earnings advanced more than 30 per cent on an annual basis, from $3149 in 1948 to $4202 in 1951. It is interesting to note that the smallest earnings gain was made by chief engineers, a job in which employment was quite stable in both years.

[19] This feature of the Boston experience will be discussed in the next section of this chapter.

[20] The influence of union policy in this development will be discussed in the next section of this chapter.

TABLE 18

AVERAGE ANNUAL EARNINGS OF BOSTON FISHERMEN BY NUMBER OF DAYS WORKED AND AVERAGE NUMBER OF DAYS WORKED, BY OCCUPATION, 1951 AND 1948

| | Average number of days worked | All fishermen | | Average annual earnings of fishermen by number of days worked | | | | | | | |
| | | Average annual earnings | Percent | 60 days or less | | 61 to 180 days | | 180 to 270 days | | 271 or more days | |
Year and occupation				Earnings	Percent of men	Earnings	Percent of men	Earnings	Percent of men	Earnings	Percent of men
1951											
All fishermen	179	$4,720	100.0	$ 528	12.9	$3,025	22.9	$ 6,110	61.9	$ 7,685	2.3
Captains	211	12,063	100.0	594	4.0	7,602	12.0	13,235	74.0	13,333	10.0
Mates	217	6,138	100.0	1,126	6.6	4,112	16.4	7,027	75.4	5,357 [a]	1.6 [a]
First engineers	219	5,621	100.0	355	4.0	2,845	10.0	6,180	82.0	6,370	4.0
Second engineers	170	4,177	100.0	502	18.5	2,928	25.9	5,970	54.4	6,709 [a]	1.2 [a]
Cooks	159	4,008	100.0	601	21.1	3,443	30.3	5,783	47.3	6,505 [a]	1.3 [a]
Deckhands	176	4,202	100.0	500	13.2	2,784	23.9	5,493	61.0	6,264	1.9
1948											
All fishermen	155	3,676	100.0	521	25.8	2,547	27.2	5,577	33.5	7,283	13.5
Captains	192	9,957	100.0	1,753	9.0	4,832	25.4	12,553	49.2	14,563	16.4
Mates	194	4,991	100.0	783	14.1	2,732	20.3	5,762	48.4	8,933	17.2
First engineers	216	5,239	100.0	523	13.2	3,462	15.1	5,429	32.1	7,335	39.6
Second engineers	162	3,692	100.0	506	23.8	2,484	27.4	5,231	29.8	7,007	19.0
Cooks	144	3,342	100.0	552	28.7	2,727	32.2	5,436	29.9	7,406	9.2
Deckhands	148	3,149	100.0	486	28.2	2,371	27.9	4,930	32.2	6,510	11.7

Source: "Annual Earnings of Boston Fishermen in 1951," U.S. Dept. of Labor, Bureau of Labor Statistics, July, 1952.
[a] One case only.

Share payments per vessel per fisherman, rather than actual earnings, show the fleet's performance in providing income under the lay. On this basis, average earnings per man day for deckhands who make up about three-quarters of the labor force rose from $21.28 in 1948 to $23.88 in 1951.[21] The first Bureau study reported that in 1948 slightly more than one-half the 49 Boston vessels studied had share payments of between $5001 and $7000; only four vessels had annual share payments of less than $4000. Similar data are available for only one other year. An earlier study covering 1933 reported annual share payments per fisherman of $494 on New England vessels of 50 tons and over.[22] Results of these two inquiries are not really comparable, for they embrace different groups, but the diversity of the findings does tend to indicate that income available per fisherman improved considerably over the span of years involved.

Available shares payment data reveal also that fishermen's earnings under the lay are highly variable from one trawler to another, from trip to trip, and from one period of the year to another. This is illustrated in Table 19, which shows trawler crew shares by trip for each of six Boston vessels from January through May, 1943. The vagaries of "fishermen's luck," differences in skill among captains and trawler crews, and movements in landing prices combine to produce the divergence in earnings. In the instance cited only the price factor may be separated conveniently. The highest shares per man on nearly all vessels occurred during the Lenten period in March and early April. Average monthly prices bid for the leading species of fish sold over the New England Fish Exchange reached their 1943 peak in March.[23] They declined before wartime price controls were imposed upon the industry in July and shares per man generally fell off from the Lenten level.

The findings of the Bureau of Labor Statistics in 1951 tend to

[21] Cf. "Summary of Information Obtained During the Investigation on Groundfish Fillets," *Report on Escape Clause Investigation under Section 7 of the Trade Agreements Extension Act of 1951*, page 151.

[22] Cf. J. R. Arnold, *The Earnings of Fishermen and Fishing Craft*, the National Recovery Administration, Industries Studies Section, Division of Review, Washington, D.C., 1936.

[23] Cf. B. E. Lindgren, "Landings and Prices of Fishery Products at Boston Fish Pier, 1943," U.S. Fish and Wildlife Service, Boston, Mass., November 1944, Page IX. The influence of Lent on fish demand will be discussed in the next chapter.

confirm the fact that there is a rather wide dispersion in earnings among vessels. For 42 vessels operating from Boston in 1951, shares per man day on an annual average basis varied from a low of $13.35 to a high of more than $32 per vessel.

TABLE 19

TRAWLER CREW SHARES FOR SIX BOSTON TRAWLERS, BY TRIP, JANUARY–MAY, 1943

Boat No. 1		Boat No. 2		Boat No. 3	
Date	Share per man	Date	Share per man	Date	Share per man
Jan. 7	$229.96	Jan. 5	$276.48	Jan. 7	$235.77
Jan. 19	436.71	Jan. 14	259.08	Jan. 19	323.96
Jan. 28	380.09	Jan. 25	345.28	Jan. 28	331.00
Feb. 8	370.11	Feb. 4	260.07	Feb. 2	295.47
Feb. 19	482.36	Feb. 15	453.91	Feb. 17	336.90
Mar. 1	278.28	Feb. 24	320.52	Feb. 26	234.31
Mar. 11	581.45	Mar. 5	223.12		
Mar. 19	448.29	Mar. 11	457.28		
Mar. 29	677.09	Mar. 24	404.23	Repair Period	
Apr. 7	450.39	Apr. 1	325.62		
Apr. 16	365.00	Apr. 12	527.16		
Apr. 23	444.35	Apr. 21	322.35	Apr. 21	429.57
May 3	288.74	Apr. 29	250.33	Apr. 29	218.00
May 11	237.53	May 10	236.09	May 10	343.90
May 20	247.96	May 18	187.90	May 18	223.04
May 28	250.18	May 26	329.35	May 26	247.79

Boat No. 4		Boat No. 5		Boat No. 6	
Date	Share per man	Date	Share per man	Date	Share per man
Jan. 14	440.36	Jan. 5	248.31	Jan. 6	454.78
Jan. 25	596.88	Jan. 12	348.89	Jan. 18	331.20
Feb. 1	99.16			Jan. 27	322.60
Feb. 9	374.54	Repair Period		Feb. 8	159.56
Feb. 17	296.35			Feb. 18	286.84
Feb. 25	202.42	Feb. 25	282.34	Mar. 1	105.57
Mar. 8	799.45	Mar. 5	408.16	Mar. 11	393.37
Mar. 15	962.22	Mar. 15	509.18	Mar. 22	471.18
Mar. 23	687.77	Mar. 24	411.29	Mar. 30	480.80
Mar. 30	593.23	Apr. 1	407.00	Apr. 8	445.32
Apr. 8	720.59	Apr. 9	615.98	Apr. 16	292.13
Apr. 15	442.48	Apr. 19	637.85	Apr. 27	274.34
Apr. 26	454.13	Apr. 28	308.50	May 7	287.27
May 5	419.40	May 7	367.51		
May 12	226.30	May 14	233.36	Repair Period	
Repair Period		May 25	308.68		

Source: Federated Fishing Boats of New England and New York, Inc.

How do the fishermen's earnings compare with those of shore-workers and those of production workers in manufacturing in Boston? There seems little doubt that on an annual basis the fishermen's earnings would be somewhat higher. On an hourly basis, Boston fishermen averaged about $1.92 during 1951; these are average straight-time earnings for a twelve-hour day. The straight-time rate of fillet cutters in Boston, about the highest paid group of shoreworkers, was $1.615; gross hourly earnings of production workers in manufacturing in Boston during the latter half of 1951 averaged about $1.54. Are the fisherman's earnings high, considering what his job entails? The fishermen would argue strongly they are not, especially when one considers the sacrifice of family life, the hard dangerous work, sometimes in heavy seas, and the discomforts of living days at a time in cramped quarters on the hazardous northwest Atlantic.

BOSTON

As Chapter II points out, large dealer-vessel owners not only employ most of this port's fishermen, but also they buy a substantial portion of the fish landed. Consequently, they have fought bitterly with the Fishermen's Union over terms of employment and the price-making process in the port. The first section below will discuss the collective bargaining experience and price determination difficulties involved. The second section will examine the price difficulties in detail.

THE BARGAINING PROCESS

The three contracts negotiated between 1939 and 1946 in Boston were signed, as mentioned above, only after aggregate losses in fishing time of more than thirteen months. The strife-torn relationship of the parties commenced in the fall of 1937. The union made the following principal demands: a guarantee of one night at home between trips; the preferential shop; the "six and six" watch which would cut the working day at sea from sixteen to twelve hours and add one man to each crew; and a guarantee of $25 per man per trip in the event of a "broker" — a trip on which the vessel failed to catch its normal complement of fish, through no fault of the fishermen.[24] Moreover, the union insisted that the dealer-vessel owners promise to stop alleged violations of the buying rules on the New England Fish Exchange, particularly the alleged practice of agreeing to split trips of fish before bidding for them on the Exchange.[25] In addition, the union demanded that the fishing captains sell their own fish on the Exchange in the place of the commission men whom dealer-vessel owners employed for that purpose.

When the companies balked at these proposals, the union conducted systematic walkouts on the vessels of each concern during the spring and summer of 1938. The owners capitulated; all the requested changes in working conditions aboard the trawlers were installed, but without a written agreement. The vessels returned to operation while the parties continued negotiations over the issue of conduct on the Exchange. Finally, on March first, 1939, the owners signed, through the Federated Fishing Boats of New England and New York, Inc., their first contract with the union.

The agreement incorporated the working conditions adopted in the previous year. It banned strikes and lockouts for its one year term, setting up a grievance procedure with a final arbitration clause for all matters except those concerning the Exchange. Both groups promised "to adhere strictly to all applicable rules and regulations of the New England Fish Exchange." [26] The parties agreed to submit their differences concerning the Exchange rules, without work stoppages, to the judge of the federal district court which had approved the Exchange rules in 1919.[27]

Both groups sought changes in the agreement when it expired in March, 1940.[28] The union asked three main alterations in the 1939 trawler lay shown in Exhibit II. First, the union requested that ice

[24] Information on the 1937–1939 experience from the boat owners, and the Atlantic Fishermen's Union; also from "The New Atlantic Fishermen's Union," *American Fisherman*, Boston, Mass., May, 1939, page 2. A "broker" is usually a mechanical failure.

[25] This is forbidden by Article VIII, *Rules and Regulations*, New England Fish Exchange. For discussion of the Exchange see Chapter III.

[26] See Section 11, Agreement, Atlantic Fishermen's Union Local #21455, AFL and members of the Federated Fishing Boats of New England and New York, Inc., March 1, 1939.

[27] For discussion of the 1919 proceedings, see Chapter III.

[28] Information on the 1940 proceedings from the boat owners and officials of the Atlantic Fishermen's Union. The latter kindly made available private papers and contracts, newspaper and magazine clippings.

costs for the summer months and one-half the cost of oil per trip be deducted directly from the gross stock instead of from the fishermen's share. Second, it asked that the companies be limited to seven in the

EXHIBIT II

THE DIESEL TRAWLER LAY IN BOSTON, MARCH 1, 1939–MARCH 15, 1940

The following items shall be deducted from the gross stock:

Wharfage	Actual cost [a]
Scales	" "
Exchange fee	" "
Chief engineer	$25.00 per trip
Second engineer	$15.00 per trip
Mate	$20.00 per trip
Sounding	$ 1.50 per day
Watching	Actual cost
Wireless operator	" "

From the crew's half of the balance of the stock the following items shall be deducted:

Fuel	Actual cost
Lubricating oil	" "
Ice	" "
Icing up	" "
Groceries and provisions	" "
Cook's per	$15.00 per trip
Water	Actual cost
Lumpers	" "

A guarantee of $25.00 will be paid to each member of the crew (excepting the captain) on all regular trips that result in brokers. . . Each regular trip shall be considered a separate venture.[b] In the event of trips that are of less than ten days duration. . . due to breakdown a daily per will be paid as follows:

Crew (maximum ten men)	$2.50 per day
Mate	4.50 per day
Chief engineer	5.00 per day
Second engineer	4.00 per day
Cook	4.00 per day

. . . The maximum payments shall not exceed the regular trip guarantees. The value of any fish that might be aboard on a trip returning due to breakdown will be credited to the following trip and all expenses in connection with the removal and sale of the fish and the other deductions from the gross stock and from the crew's share will also be carried forward to the following trip.

Source: Agreement, Atlantic Fishermen's Union, Local #21455, AFL and Members of the Federated Fishing Boats of New England and New York, Inc., March 1, 1939.
[a] "Actual cost" in the lay means cost to vessel owner.
[b] This section is presented here in summary.

number of lumpers, or "fish longshoremen," whose costs the concerns might deduct from the fishermen's share. Third, the union insisted that it name the supply houses from which trip provisions, paid for from the fishermen's share, should be bought by the vessel owner.

The New England Fish Exchange rules continued to be an issue. The parties had been unable to resolve their differences through the machinery set up in the previous contract. Consequently, the union demanded representation on the board of directors of the Exchange along with the dealers. Moreover, the union continued to insist that the fishing captains, not commission men, auction all fish.

The dealer-vessel owners not only turned a deaf ear to these proposals, but also they made the important demand that one apprentice per boat, whom they would pay for six months, should then be allowed to join the union. Negotiations broke down quickly. The owners tied up their vessels. The ensuing 102-day stoppage paralyzed the port until June, costing the industry at least three million dollars in lost production during the most favorable fishing season of the year. Too exhausted to fight longer, the parties signed a new agreement extending to March 15, 1942. It recognized the ice cost changes sought by the union, provided for arbitration of the cost whenever the number of lumpers hired should exceed the number seven, and stipulated that the union should admit novices "under reasonable rules and regulations." [29] The agreement maintained the *status quo ante* on virtually all other issues.

Despite the bitterness which surrounded the adoption of this agreement, the parties amended it successfully in December, 1940. The Federated members agreed to add two or three men to their trawler crews to alleviate unemployment among the port's fishermen caused by the sale of nine vessels to the United States Navy for defense duty.[30] In return, members of the union agreed to save fish livers and other saleable products formerly discarded at sea. The proceeds from sales of these items were to be included in the gross stock.

Hostilities broke out next on January 1, 1942. The union insisted

<hr>

[29] Cf. Agreement, The Atlantic Fishermen's Union, Local #21455, AFL and members of the Federated Fishing Boats of New England and New York, Inc., June 21, 1940. (Quotation is from section 4 of the contract.)

[30] Cf. "Navy Secures Hub Trawlers," the *Boston Post*, August 15, 1940.

that the dealer-owners pay the full cost of war risk insurance for crew members. The companies declined. The Federated vessels did not sail again until February 13, when, with reluctance, their owners complied with a National War Labor Board order to return their vessels to operation.[31] The Board settled the issue later on April 3 by ordering the parties to divide the insurance cost evenly.[32]

Meanwhile, in March, the union had sought to replace the expiring 1940 agreement with a new contract providing for company-paid war risk insurance and a bonus of $20 per man for every trip begun while the war continued.[33] Negotiations proved fruitless. The vessels continued to sail under the terms of the 1940 agreement as amended in April by the War Labor Board award.

Difficulties which were to last for the duration of the war cropped up in July, 1943 as the result of ceiling prices imposed upon fish by the Office of Price Administration.[34] The maximum price order established producer ceilings at the level of the weighted average price prevailing for each species during 1942. This action "rolled back" prices between 10 per cent and 25 per cent from their "free level" on July 1, 1943, but it left most prices substantially above their 1940–1942 average. The fishermen insisted that this "rollback" cut their wages as much as 36 per cent. Crew members tied up the fleet for two weeks while their leaders argued with OPA for higher prices. The price agency refused adjustments; the men returned to work, in compliance with a War Labor Board order, to "try" the new prices for a short period.[35] The experiment did not satisfy the

---

[31] Cf. Press Release, National War Labor Board, P.M. 2471, February 11, 1942.

[32] Cf. *In the Matter of Federated Fishing Boats of New England and New York, Inc. and the Atlantic Fishermen's Union, SIU, AFL*, Case No. 16, National War Labor Board, April 3, 1942.

[33] Information from proposed contract of March 15, 1942, held in files of the Atlantic Fishermen's Union.

[34] Cf. Office of Price Administration, "Maximum Price Regulation 418," set forth in *Fishery Market News*, U.S. Fish and Wildlife Service, Washington, D.C. March, 1944 Supplement.

[35] The New England Regional War Labor Board ordered the men and vessel owners to return to operations immediately. It did not assume jurisdiction in the dispute because it felt "the issue . . . is essentially an issue of the price to be paid for fish rather than a wage or salary issue . . . The New England Regional War Labor Board does not possess authority to fix prices at which fish handled by the operators can be sold." (See *In the Matter of Federated Fishing Boats etc. and the Atlantic Fishermen's Union, SIU, AFL*, Case No.

fishermen. They struck again from November 19, 1943 to the middle of January, 1944 to force a price change, but without success.

Before proceeding, we should note that the procedure of the OPA in establishing ceiling fish prices provided one important reason for Boston's troubles. The agency set the same levels for all New England ports. Historically, however, Boston prices had led those in the other centers, in some cases by several cents.[36] Consequently, the ceilings cut sharply most Boston prices prevailing in the summer of 1943. In the other ports, the ceilings represented but a small drop, and in some cases, an increase in fish prices. Efforts of Boston interests to secure a differential which would reflect historical price differences among the ports met with strong opposition from Gloucester, New Bedford, and Portland. The OPA took no action.

Balked in its quest for higher price ceilings, the union asked the Federated to make the following principal changes in the lay in February, 1944: [37]

a. Make the cost of fuel oil, lubricating oil, and ice deductible from the gross stock instead of from the fishermen's share of the catch.

b. Compute the cost of fuel oil, lubricating oil, food, and ice at 1942 prices instead of actual cost, the difference to be paid by the owners.

c. Eliminate from the lay all other deductions from gross stock, including the cost of lumpers.

d. Extend the new lay for two years beyond the termination of the war, with the guarantee that the fishermen's wages be determined at fish prices no less than the wartime ceilings set by the Office of Price Administration.

In addition, the union requested that there be established a sick and death benefit fund, jointly financed by the fishermen and vessel owners. Finally, the union asked that the fishermen receive two trips vacation per year with pay.

111–1917–D, July, 1943.) The Atlantic Fishermen's Union kindly made available all papers in this case from the union's files.

[36] Cf. *Report of the Federal Trade Commission on Distribution Methods and Costs*, Part VIII, June 30, 1945, pp. 33–40 and pp. 66–68.

[37] Cf. *Federated Fishing Boats of New England and New York, Inc. and the Atlantic Fishermen's Union, SIU, AFL*, National War Labor Board, Case No. 111–7252–D, 1944 and 1945. I am indebted to the Atlantic Fishermen's Union for making available the entire record in this case from the union's files. All facts on the Boston 1944–45 experience are drawn from this record.

The owners rejected all demands. Between March 30, 1944 and August 1, 1945, the case received full hearings and review before all levels of the National War Labor Board. Finally, the report of the latter's review committee rejected nearly all proposals.[38] The committee recommended only that fees for wharfage, scales, use of the New England Fish Exchange, and watchman's expenses — "above-the-line expenses" — be eliminated from the lay on the ground that the latter charges occurred after the fishermen's work had been completed and were really "managerial" in nature.[39] Figures introduced by the Federated indicated sharp increases from prewar levels in the average share per trip per fisherman as Table 20 demonstrates. Consequently, the committee refused to recommend changing the general structure of the lay and the use of 1942 prices in computing cost deductions because these moves would give the fishermen an unjustified wage increase. The committee felt that extending the lay for two years with guaranteed landing prices to the fishermen would run contrary to the essential profit- and risk-sharing nature of the lay. Furthermore, it would place impossible economic restrictions on the industry. The committee could not agree on a vacation plan, and denied the two-trip per year vacation arrangement requested by the union.

Thoroughly dissatisfied with the War Labor Board experience, the union changed tactics in the fall of 1945. It asked that a "60–40" lay arrangement, which would give the crew members 60 per cent of the net stock, replace the standard "50–50" split on a few trawlers returning to the fleet after war service. In December, the entire Federated membership tied up its vessels in support of the few

[38] *Ibid.*, "Proposed Report of Review Committee to National War Labor Board," July 25, 1945. The labor member dissented as follows: he favored guaranteeing fish prices (d. above) and argued that stabilized prices in the postwar period would benefit both parties; he favored using 1942 prices for supplies (b. above) because he believed the OPA had restrained fish prices more successfully than the prices of goods used by the fishermen; he favored establishment of a contributory sick and death benefit program; he favored altering the lay and reallocating fuel, ice, and other costs as requested by the union (a. above) because they are necessary to conduct of the shipowner's business and hence one of his costs; and he favored the vacation proposal as consistent with the Board's vacation policy recommendations in other cases.

[39] The industry member dissented. He insisted these charges were not peculiarly managerial. He maintained that the proposed change would have "unstabilizing" effects.

owners who refused to grant the new lay division. The ensuing stoppage lasted six months; it ended with the installation of the "60–40" on all trawlers, a complete victory for the union.[40]

TABLE 20

BOSTON AVERAGE SHARE PER FISHERMAN, 1940–1945, DATA SUBMITTED TO THE
NATIONAL WAR LABOR BOARD BY THE FEDERATED FISHING BOATS OF
NEW ENGLAND AND NEW YORK, INC.

Year	Average annual share	Average share per trip	Share per trip increase over 1940	Percentage increase over 1940
1940	$2500[a]	$ 83.33	. . .	. . .
1942	5303.90	174.76	$ 91.43	110
1943	8831.92	287.68	204.35	245
1944	7479.03	261.05	177.72	213
April 30, 1945	. . .	279.67	196.34	236

Source: "Informational and Statistical Data, Federated Exhibit No. 1," Table 8, *Federated Fishing Boats of New England and New York Inc. and Atlantic Fishermen's Union, S.I.U., A.F. of L.*, National War Board, Case No. 111–7252–D, 1944 and 1945.

[a] Included in the document is the following note: "The parties agree that the annual share in 1940 is $2500. The average number of trips per year is 30. The agreed annual share in 1940 of $2500 divided by thirty trips is an average share per trip of $83.33."

Sheer bargaining power determined the issue. The union wisely encouraged full operation of the Gloucester and New Bedford fleets.[41] Nonstrikers in those ports made available temporary sites [42] which eased the economic pressure on the Boston fishermen. The Boston dealer-owners, on the other hand, became increasingly desperate for

[40] Cf. "Uniform Trawler Agreement," Atlantic Fishermen's Union, SIU, AFL and members of the Federated Fishing Boats of New England and New York, Inc., May 15, 1946. In addition, the union won an increase in "broker" payments to $5 per day per fisherman up to a maximum of 10 days. The General Seafoods Corporation held out nearly a month longer than the rest of the Federated, but finally capitulated on all points. We should note that Saul Wallen, Chairman of the New England Regional War Labor Board, told the author that he drew up a "model" contract for the parties in the latter part of 1945 in the hope of compromising their differences, but to no avail.

[41] This was in accord with policy implied in the Union's Constitution. Article XI, "Strikes and Collective Bargaining," Section B, stipulates, "strikes at any port while there is a strike in another port in the same geographical area shall be called only by general ballot." (*Constitution and By-laws*, Atlantic Fishermen's Union, SIU, AFL, March 1, 1944.)

[42] A "site" is a job aboard ship.

fish supplies to maintain their markets against Canadian, Newfoundland, and Iceland competitors who were making an unprecedented bid
for American patronage.[43]

Assuming the same cost and price conditions as before the change,
the new lay split gave the Boston fishermen a raise of about 25 per
cent at the expense of the owner in sharing any given net stock. For
example, a $10,000 net stock with $1000 in "below-the-line expenses"
under the new arrangement would yield each fisherman in a seventeen-
man crew about $294, or approximately 25 per cent more than under
the old division.

The 1946 agreement was meant to "mark the beginning of a new
era"[44] in which the union and the owners were to forgive and forget
past misunderstandings and grievances; both groups were to coöperate
to insure the industry's prosperity.

Unfortunately, no such transformation occurred. When the agreement expired in 1947, the parties were more bitterly at odds than
ever; both wished a new contract, but they could not even begin
negotiations for one. They continued operations without an agreement, but largely in accord with the 1946 arrangement.

In summary, the record reviewed in this section shows that union
policy, in bargaining with Boston dealer-owners, sought (1) more
favorable lay arrangements and (2) some control of the Boston
price-making mechanism. The dealer-owners balked successfully the
latter union aims, but only at the cost of substantial lay concessions.
Despite these, the price determination process remains the focal point
of union clashes with dealer-owners. The major stumbling block is
the "sellover" feature of the Boston price-making mechanism, which
we shall now discuss.

THE "SELLOVER"

Continued controversies over the process of price determination
on the New England Fish Exchange have prevented the development
of harmonious relations between Boston fishermen and dealer-vessel

---

[43] The problem of foreign competition is discussed in the next chapter.
Imports of frozen fillets rose more than 350 per cent between 1939 and 1945,
however.

[44] See Paragraph 27, "Uniform Trawler Agreement," May 15, 1946.

owners. The fishermen have argued that the dealers violate the rules of the Exchange daily in their zeal to obtain fish at low prices. The fishermen have rebelled particularly against the "sellover," or fish resale provision of the Exchange rules. As is pointed out in Chapter III,[45] this arrangement allows a dealer to refuse to take any more fish after a certain quantity has been removed from the vessel, if he feels that the fish are not first quality. If the dealer is successful in his stand, the fish remaining in the vessel must be "sold over" or auctioned a second time. All buyers, including the complainant, are eligible to bid at the sellover. The resale nearly always results in lower prices for the remainder of the fish on the vessel.

In practice, the fishermen usually challenge the dealer's quality protest and maintain that the fish are still "No. 1" quality. Under the Exchange rules, an arbitration board consisting of the dealer, the fishermen's representative, and any mutually acceptable third person willing to serve is then invoked to settle the dispute. Most often the dealer and the fishermen have not been able to agree on a third party; under the rules, the Exchange manager then must appoint the third person, or decide the issue himself as a practical matter. The fishermen have lacked faith in the Exchange manager's decision. He is indirectly an employee of the dealers; the latter make up the Exchange membership and are represented upon the Exchange board of directors. Moreover, the present manager, who has held the position for more than twenty-five years, has also operated a supply house which sells materials to the dealers.[46]

The union charges that the "sellover" is a price-cutting device "built into the market mechanism" to depress the fishermen's wages. Union representatives point to the vague quality classifications of "No. 1" and "No. 2" in the Exchange rules.[47] Further, union officials insist that the sellover merely allows the dealers to evade responsi-

---

[45] Section on Boston.

[46] From his limited acquaintance with Mr. E. L. Dunn who has been President of the Exchange for over a quarter of a century, the writer would not hesitate to say that Mr. Dunn impresses him as an honest man. But that unfortunately is not the problem. The difficulty is that the system will not work when the key man is in so many ways involved with one of the parties. As we shall indicate later, the function of determining quality disputes simply should not be in the hands of the Exchange manager, no matter who holds that position.

[47] See Chapter III.

bility for their original bids. Finally, union spokesmen allege that all fish bought by the dealers, whether "No. 1" or "No. 2," are comingled, processed, and sold to secondary wholesalers without differentiation of quality or price.

The dealers not only deny vociferously the union's contentions, but also they claim that the fishermen's incomes would be lower if the sellover were eliminated. The dealers insist that the sellover feature makes it possible to bid extra amounts for the higher quality fish. They consider the sellover essential to the maintenance of their competitive position.

Where does the truth lie in this welter of charges and counter charges? To find the answer let us look at the way in which the "sellover" works.

The fishermen and vessel owners clash on the Boston market because the former are interested in obtaining maximum prices for their catch in order to make the best possible income per trip under the lay system. Nearly all the larger vessel owners, on the other hand, are also fish dealers and must keep a "weather eye" on the price of fish, for as dealers, raw fish prices are their most important cost item. With the exception of the chain stores, which do not own vessels, the dealer-owners are the largest fish purchasers on the New England Fish Exchange.[48]

To set the picture, imagine for the moment that on a given day there are ten vessels at the Pier with about a million pounds of fish. In that catch, there will be several species. For the sake of simplicity, however, assume that there is only one species and that the dealers bid at open auction for the supplies available.[49]

The only restriction upon the dealer-owners as the bidding proceeds will be that they may not bid on fish landed by their own vessels. (The former is forbidden by the Rules of the New England Fish Exchange. They allow a dealer to take fish from his own vessel only when a satisfactory bid from other is lacking; as proof of the latter, the dealer taking his own vessel's fish must register the price paid for them with the Exchange office.)[50] Hence, the dealer-owners

---

[48] The A & P and the First National Stores, Inc. are the chain stores which have buying and processing operations at the pier.

[49] See Chapter III.

[50] Cf. the New England Fish Exchange, *Rules and Regulations*, Article V.

will be bidding for fish from vessels owned by other dealers. Providing the major dealers (most of whom, with the exception of the chain stores, we stress again, are also vessel owners) do not conspire to keep prices down, the prices which result from the auction should approach the "competitive ideal." The union has charged often that there is such collusion, but has not succeeded in supporting the charge with formal evidence.

When the bidding has ended and the fishermen and "lumpers" have unloaded one thousand pounds or more of the fish from a given vessel, the "sellover" may enter the scene. Let us suppose, then, that on the day in question, after some 10,000 pounds have been removed from a given vessel, the buying dealer-owner calls a halt and insists that the fish are no longer "No. 1" or first quality; he points out that in bidding in the first sales market he, like the other buyers, proceeded on the assumption that all fish for sale were "No. 1."

Since there are no explicit standards set by the Exchange Rules for objectively distinguishing first from second quality, and since too the "sellover" or second sale almost always results in a lower price for the balance of the catch, the vituperation will be long and loud, for the fishermen will oppose the move bitterly. One other development may "baffle" the fishermen further. The dealer-owner whose fish are being sold may side with the buyer. Why should he do this? As the vessel owner, should he not be as interested in maximizing fish prices as is the crew?

The answer is that he may not be; he has to split the net stock 40–60 [51] with the fishermen (the latter getting the larger share) every cent received as a vessel owner. On the other hand, every cent he saves as a buyer of fish is his own. It is true that in the situation we have posited, he is acting in the capacity of vessel owner, not as a dealer. But on that very day or on some other day soon, the tables will be turned; he will be in the buyer's shoes and the very dealer who, in our example, is currently the buyer and seeking a "sellover," may be the vessel owner. In view of these realities, the vessel owner would seem to be acting in his own interest in siding with the buyer. No overt collusion is necessary. The buyers are simply facing the fact of life, under the current market setup, that opposing one another

[51] For definition of this term see first section of this chapter.

can work only to the fishermen's benefit. We have already explained that the fishermen consider the machinery for settling the quality issue to be unfair.[52] How could such a situation as this help but be the setting for a first class melee?

As if the situation as we have described it thus far were not volatile enough, one other factor enters the picture. Suppose for example, that the fish which the buyer currently is questioning were to be about the same as those the fishermen had landed on a previous trip when no one raised any demand for a "sellover." Such a set of circumstances would be certain to confirm the fishermen's judgment that the "sellover" is a "racket," or to put it in their words, "the tolling of the bell" to announce a "sellover" is really the ringing of the "burglar alarm"! [53] The fact is, however, that the operation of the sellover under the present *Exchange Rules* is at least as dependent upon the state of the market as upon the physical qualities of the fish.[54] The dealers are most likely to question quality when demand is only fair and supplies of fish are large. It appears, for example, that the same quality fish may be "top" quality on Friday when supply is short in relation to demand and second quality on Monday when the opposite market conditions prevail. The dealers seldom requested the "sellover" during the second World War.[55] That experience strengthened the fishermen's belief that the "sellover" is simply a price-cutting device.

The operation of the "sellover," then, depends on both fluctuating market factors and the issue of quality. This confusion of forces at work, when considered with the objections of the fishermen to the whole machinery for quality determination and the tendency for the dealer vessel owners to be somewhat in sympathy with one another because they are aware of the importance of primary market prices in determining processing costs, yields a system almost preordained to cause an economic war.

As a device for promoting emphasis on quality in highly perishable

---

[52] See page 76.

[53] Cf. Henry Wise, Esquire, "Brief for the Appellants," *Patrick J. McHugh et al. v. Commonwealth of Massachusetts*, Equity No. 4897, Supreme Judicial Court of Massachusetts, February, 1950, page 12.

[54] See Appendix.

[55] The Office of Price Administration made no attempt to grade fish for pricing purposes.

food operations, the sellover, properly conceived and executed, might be desirable in the Boston market. But changes must be made to eliminate the issues which the process heretofore has raised. Sellover disputes have also interfered with coöperation between Boston's fishermen and dealers on most other matters. Since mid-1949, for example, the Fishermen's Union has been urging the return of some of the twelve trawlers sold to the United States Army early in 1949 for use in the German North Sea fisheries. The union argued that the Army has found the trawlers unsuited for North Sea duty and that they should be returned to increase Boston's catch.[56] The dealer-owners retorted that more vessels are not needed. Their solution is fuller operation of the present fleet, with less shore time for fishermen between trips. The fishermen in return have stated that they see no reason for making more frequent trips so long as the dealers refuse to consider changes in the sellover rules. Discussions have bogged down in argument at that point.

The fishermen's solution to the problems raised by the sellover is to wipe out what they call "this tool of exploitation." Is this the only or the best answer? Chapter VII will suggest that it is not.

Gloucester and New Bedford

Collective bargaining between the Atlantic Fishermen's Union and the vessel owners in Gloucester and New Bedford has worked smoothly because most of the latter are not fish dealers. Primary market price disputes between the fishermen and dealers in these ports are outside the bargaining process; the Fishermen's Union deals with the clearcut business unit of the fishing vessel. Both fishermen and vessel owners have the basic common interest of maximizing the value of the catch. Both have reasonably adequate information about the economic limitations of the other. Many owners are active

[56] The Union's argument concerning the trawlers' German experience is at least partly correct. The trawlers operated profitably during the herring season, but German interests refused to buy them at the prices paid by the Army to U.S. owners. Moreover, they voiced no objection to returning the trawlers to the U.S. See "American Trawlers in Operation in German Fisheries," *Fishery Products Report No. 180*, U.S. Fish and Wildlife Service, Boston, Massachusetts, September 15, 1949; also "Operation of American Trawlers in Germany," *Fishery Products Report No. 38*, U.S. Fish and Wildlife Service, Boston, Massachusetts, February 24, 1950. Please note Postscript to this book. Some of these trawlers are back (1953) but will operate from Rockland, Maine.

or former fishing captains; they have some understanding of the fishermen's attitudes and problems. The fishermen and their union representatives, on the other hand, are well acquainted with the economics of vessel operations. Both ports have several lays drawn to meet the circumstances of different sized vessels and varied types of fishing. Sample lays are shown in Exhibits III, IV, and V.

EXHIBIT III

THE "CLEAR 40" DRAGGER LAY FOR GLOUCESTER AND NEW BEDFORD

(a) There first shall be deducted from the gross stock the following items:

Chief engineer ($10.00 per trip redfishing,[a] $15.00 per trip when cod or haddock fishing.)

Second engineer ($5.00 per trip.) [b]

Fathometer ($1.50 per day except when under overhaul or repair.)

Lumpers (All those not hired by the crew.)

Rags, watch, light bulbs for forecastle, deck, and engine rooms only, replacement of dishes.[c]

Ice from June 1st to September 30.

(b) From the crew's said 60 per cent of the remaining stock there shall be deducted the following items:

1% of said 60% for union benefit and welfare fund [d] when or once requested by the union.

Food

Cook ($10.00 per trip.)

Mate ($5.00 per trip, unless another "per" prevails.) [e]

Fuel oil

Lubrication oil

Ice from October 1st to May 30.

(c) From the owner's said 40%, 10% thereof shall be deducted and paid to the captain.[f]

Source: "Working Agreement of Gloucester 60–40 Draggers," 1948; "Agreement Atlantic Fishermen's Union and members of the New Bedford Seafood Producers' Association," April, 1946.

[a] Not in New Bedford lay; no redfishing from New Bedford.
[b] $10.00 per trip in the New Bedford lay.
[c] Not specified in the New Bedford lay.
[d] Not specified by union up to June, 1950.
[e] $10.00 "per" in New Bedford and *deducted from gross stock.*
[f] Not specified in New Bedford lay.

Generally, the more liberal terms for fishermen appear in the smaller vessel lays because the owners' equity in boats and equipment is less on those classes.

The lay arrangements are substantially the same as in pre-union days. Only the large, or "clear 40" dragger lay has changed materially under union pressure. After a 1940 strike of ten weeks in New Bedford, the union forced dragger owners to make ice costs for the summer months a deduction from gross stock instead of the fisher-

EXHIBIT IV

GLOUCESTER GILL NET LAY AND NEW BEDFORD SCALLOPER LAY

Gloucester Gill Net Lay [a]

To be deducted from the gross stock:

One-half (½) of 1% for Master's Producers Ass'n.
Fuel oil
Gasoline
Gear oil
The engineer shall receive $1.00 for each day the boat sails.

After deduction from gross stock the boat shall take 50%. From the boat's share of 50% a deduction shall be made for the captain's percentage.

To be deducted from crew's 50%:

Groceries
Cook's wages — $1.00 each day worked on board.
Reeling

New Bedford Scalloper Lay [b]

To be deducted from the gross stock:

Watchman when the scallops or the groceries for the trip are on board.
Scallop bags.
Replacement of broken dishes.
Sounding machine (if any) $ 1.50 per day when in use.
Engineer $15.00 per trip.
Cook's percentage $10.00 per trip.

After deductions from the gross stock the crew shall take 65% of the remaining stock.

To be deducted from the crew's share of 65%:

Groceries
Coal
Wood
Fuel Oil
Ice

[a] "Gloucester Gill Net Lay," 1947.
[b] "Agreement Atlantic Fishermen's Union and New Bedford Seafood Producers' Association," April, 1946.

men's share.[57] The union secured the same concession from Gloucester operators in 1945 as an award of the Regional War Labor Board.[58] In both ports, the union has insisted successfully that the fishermen be paid an additional hourly or daily wage for any repair work performed on the vessel while it is in dock.

EXHIBIT V

THE "BROKEN 40" LAY

The lay of the boat shall be as follows: [a]

To be deducted from gross stock:

Ice
Oil
Cook
Engineer
Lumpers
Coal and wood

After deductions from gross stock, the boat shall take 40% of the remaining stock.

To be deducted from the crew's share of 60%:

Groceries

Sources: "Agreement of Atlantic Fishermen's Union and the members of the New Bedford Seafood Producers' Association," April, 1946. Mr. Melvin I. Bernstein, Secretary of the Gloucester Vessel Owners' Association, Inc. stated that the Gloucester "broken 40" lay contained the same items similarly arranged. He added fathometer cost as a deduction from gross stock.

[a] Applies to small draggers.

The collective agreement has benefited the Gloucester and New Bedford fishermen by providing job protection, formalizing the fisherman-owner relationship, and, to some extent, by making provisions for safety protection aboard the vessels. Agreements have stipulated the employment of union fishermen only, so far as is possible; [59] they have established formal grievance procedures with final arbitration clauses; they have required that a clear statement of earnings and expense deductions be furnished each fisherman; they have specified that itemized bills covering all expense deductions

[57] Cf. "Local Owners and Union Sign Contracts," the *New Bedford Standard-Times*, April 13, 1940.

[58] Cf. *In the Matter of Gloucester Fishing Vessel Owners' Association and the Atlantic Fishermen's Union, SIU, AFL*, Case No. 111–15979–D, August 31, 1945. The Atlantic Fishermen's Union kindly made this case record available from the union's files.

[59] Up to 1947 agreements contained preferential shop clauses.

from the fishermen's share be filed with the union. Although the union has concentrated its safety improvement efforts on a program to achieve federal inspection of fishing vessels,[60] the union's agreements require medical supplies aboard each boat, and in some cases, extra life rafts. Safety at sea has been a major union problem in Gloucester and New Bedford because of the large number of wooden and smaller vessels. In New Bedford, for example, 33 men lost their lives through four ship sinkings between July, 1949 and April, 1950.[61]

Gloucester and New Bedford vessel owners have watched with interest the efforts of the union, discussed in the next section of this chapter, to increase the price of fish landed. The union's policies have often limited output and reduced the number of trips, but the owners have made few efforts to interfere.[62] Since 1941, when the Federal Trade Commission ordered Gloucester mackerel boat owners and the union to cease efforts to fix prices through joint production controls,[63] the operators have carefully avoided coöperation with the union on such matters.

In the face of the Labor-Management Relations Act of 1947, which prohibited the preferential shop clause,[64] New Bedford owners and the union did not negotiate a contract to replace that running out in 1947 until 1952. In the meantime, both sides abided by the old agreement.

A federal antitrust indictment brought matters to a head in June of the latter year. A federal grand jury charged the Fishermen's Union and the Seafood Producers' Association separately with having conspired to restrain production and unlawfully raise prices. So far as the indictment concerned the fishermen, it rested on the premise that

[60] The union has sought since 1941 to bring fishing vessels under federal inspection laws. See testimony of Patrick J. McHugh in "Unofficial Public Hearing on Marine Inspection for Fishing Vessels," Municipal Building, New Bedford, Massachusetts, April 17, 1950 (Mimeographed by New Bedford Seafood Producers' Association, New Bedford, Massachusetts).

[61] Information from the New Bedford Seafood Producers' Association. See also, "New Demands for Inspection Follow Latest Trawler Losses," *Quincy Patriot Ledger*, April 11, 1950.

[62] See, however, "Boatowners Term Catch Quotas Illegal, Ask Repeal," *New Bedford Standard-Times*, January 25, 1947.

[63] Cf. *Report of the Federal Trade Commission on Distribution Methods and Costs*, Part VIII, June 30, 1945, pp. 60–63.

[64] Cf. "Labor-Management Relations Act of 1947" (N.Y.: Prentice Hall, Inc., 1947), Section 8.

"The fishermen who are members of the Atlantic Fishermen's Union are not employees, workers, or laborers, but are independent business men engaged in fishing for their own account and profit . . . Each fisherman who is a member of the crew risks making a profit or a loss, according to the success or failure of each fishing trip . . . "[65] Insisting that they are employees, the fishermen struck for three weeks and brought the owners to add the following major provisions in a new two-year contract:

a. The language "employees" is set forth specifically and the owners acknowledge such to be the fishermen's status.
b. The owners agreed to a "union shop."
c. The owners agreed to pay $\frac{1}{2}$ of 1 per cent of the gross stock of each trip into a welfare fund to be established and managed jointly by the Association and the Union.
d. The owners agreed to "broker" payments of $1 per day.
e. Four days' rest ashore between scalloping trips and three days' rest between dragger trips was specified.
f. Provision is made for reopening of the agreement should any part of it be invalidated by the antitrust proceeding.

While the union was anxious to secure the benefits outlined for the sake of the members, it was even more insistent upon them in order to point up the fact that the fishermen are "employees," not "independent businessmen." It remains to be seen how the court will decide the question of whether the fishermen are "employees" or something else. It remains to be seen, too, what kind of view the court will take of the collectively bargained provisions for several days rest between fishing trips. As of the fall of 1952 the case had not come to trial, so that the only definite consequence of the antitrust proceedings in that year had been to stir the parties to bargain a contract — something they had been unable to achieve in the preceding three years.

Gloucester vessel owners and the union adopted a new agreement in 1948.[66] It met the union security problem — which had held up

---

[65] Quote is from document on file at office of the Atlantic Fishermen's Union, Boston, Mass.

[66] Cf. "Working Agreement of Gloucester 60–40 Draggers," Agreement of the Atlantic Fishermen's Union, SIU, AFL and the Gloucester owners individually, 1948.

negotiations for a prolonged period — by providing that the owners would give preference to "experienced fishermen"; the latter were required by the contract to have had two to five years practical fishing behind them. Virtually all such people in the port are union members. In addition, the agreement gave the union the right to establish other reasonable standards of ability and experience from time to time.

The Atlantic Fishermen's Union set up the selling rooms in Gloucester and New Bedford to promote competition among the fish dealers and to eliminate controversies over determination of the value of fish landed. The mechanics of these auctions are described in Chapter III.[67]

In both selling rooms, each fishing captain sells his own catch. Each is free to choose the most satisfactory bid, offered by the competing buyers to the union representative as intermediary, or each captain may refuse all bids and offer the catch at a later sale.

The auctions have substituted open bidding for private deals between fishing captains and dealers. They have lessened the opportunity for collusion among buyers and have given the union a record of prices established. They have not stopped controversies over the proper value of fish landed.

At Gloucester, disputes have centered about weight allowances and weighing methods. The weight allowance issue involves determination of the proper percentage to be deducted for ice and trash in setting finally the value of a given cargo of fish. The union has continued to insist that no more than 5 per cent be subtracted for this purpose from the total trip weight. The dealers have sought to deduct up to 20 per cent.

The union sought to settle the question in 1947 by barring from bidding in the selling room a dealer who violated the union 5 per cent stipulation; the Massachusetts' courts action in forbidding this technique will be examined in the last section of this chapter. The union next forced dealers to use de-icing machines in unloading the catch. When arguments continued, the dealers agreed to submit samples, as the vessel was unloaded, to a joint committee representing the crew and the dealer. The committee was to determine a fair weight deduction for ice and trash, but controversies persisted.

[67] See sections on Gloucester and New Bedford.

Differences over weighing methods have intensified bad feeling. The fishermen maintain that some dealers' scales weigh "short." In May, 1949, an arbitrator suggested the use of bonded weighers. The union secured them, but the dealers refused to buy fish tallied by the bondsmen. They purchased every trip with the limitation "no bonded weigher" as part of the bid.[68]

Similar weight difficulties have plagued the port of New Bedford. In the winter of 1947, for example, the fishermen tied up in protest against weighing methods. They insisted that wire baskets be used in unloading to secure true weight for fish landed. They returned to work after the dealers consented to use the baskets for one month.[69] After the trial period, uniform use of the baskets did not continue; weighing methods have remained a source of periodic work interruptions.

Moreover, some New Bedford dealers have tried to lower prices by several cents per pound below what they had previously bid in the selling room on the ground that not all fish were first quality. To meet that problem, the union has employed a sales contract which the dealer and captain must sign in the selling room. Nevertheless, at least one dealer has attempted to cut prices at his dock, even after signing the selling room contract.

At both ports these persistent skirmishes have tied up the boats and delayed sailings intermittently. Results have been costly. In the first instance, the fish have been left to deteriorate in the vessel while the parties argued; in the second the number of fishing trips has been reduced.

Thus, the union-run market mechanisms in Gloucester and New Bedford, like the dealer-operated Boston auction, have not provided satisfactory machinery for determining catch values. The absence, in Gloucester and New Bedford, of the sellover feature and the integration of fish dealing and trawler owning, unlike the case in Boston's price determination troubles, has not made the process of catch value determination any smoother in the former ports. How might the

[68] Cf. "Rules Regarding Weighing Out of Trips at Gloucester," U.S. Fish and Wildlife Service, *Boston Fishery Products Report*, Boston, Mass., June 10, 1949, page 4.

[69] Cf. "Dealers Consent to Month's Use of Wire Baskets," the *New Bedford Standard-Times*, February 25, 1947.

market systems in Gloucester and New Bedford be altered to improve their performance? Chapter VII will explore the possibilities.

Job Control

Strong union security clauses in Gloucester, New Bedford, and Boston contracts [70] preserve available job opportunities for members of the Atlantic Fishermen's Union. One of the principal duties of union officers is to see that no vessel leaves port unless it has a full union crew.[71]

During the second World War, the union tightened its admission requirements to control more effectively the number of union men eligible for employment. The membership voted to raise the initiation fee from $10 to $100. New constitutional provisions adopted in 1944 stipulated that all applicants be approved by vote of two out of three regular meetings in the port concerned.[72] New entrants are on probation for six months; after that they gain full membership if not barred by vote of the port's regular meeting. Needless to say, the growing infrequency of regular meetings, pointed out in Chapter IV,[73] has made the membership approval requirement a major stumbling block to admission.

Moreover, the union has made temporary exit from and readmission to its ranks much less convenient. The 1944 changes forbade the issuance of retiring cards. Any fisherman whose dues payments fall over six months in arrears is removed from the membership rolls. Reinstatement requires a favorable vote by members in the port at two out of three regular meetings and compliance with "such terms and conditions as the union sees fit to impose." [74] As a minimum, the "terms and conditions" usually provide for payment of all back dues and a $50 fine.

[70] Up to 1947 contracts had preferential shop clauses requiring that union fishermen be hired ahead of non-union men. Union shop clauses negotiated since 1947 also have called for experienced fishermen — in effect union men.

[71] Cf. "Duties of Officers," Article VIII, *Constitution and By-laws*, Atlantic Fishermen's Union, SIU, AFL, March 1, 1944. This provision has appeared in the union constitution from the beginning.

[72] Cf. "Membership," Article III, *Constitution and By-laws*, Atlantic Fishermen's Union, SIU, AFL, March 1, 1944.

[73] Section on the Atlantic Fishermen's Union.

[74] Cf. "Income and Dues," Article XIV, *Constitution and By-laws*, Atlantic Fishermen's Union, SIU, AFL, March 1, 1944.

It is difficult to appraise the effectiveness of the union's entry limitations and their effect on the industry. The organization does not keep annual total membership records. It maintains merely a total of quarterly dues payments. These figures are not satisfactory for estimating the number on the rolls, for the fishermen often pay several months' dues at a time.

The available evidence indicates that the limitation policy has been effective, at least in Boston. It has not, however, created any scarcity of candidates for available jobs. Total membership in 1939 averaged about 2900. It increased irregularly between that year and 1947, when it reached an average of about 3760.[75] Discussion with leaders of the union indicates that most of the increase in these years took place up to 1945 and was associated with the pronounced expansion of fishing activity in Gloucester and New Bedford. Between 1947 and 1951 membership dropped to about 3300.[76]

It is likely that most of this decline took place in Boston. The Bureau of Labor Statistics wage studies analyzed in an earlier section [77] found that between 1948 and 1951 the number of fishermen finding employment on the Boston Fish Pier fleet (virtually all union members) declined from 1416 to 1092, while the maximum number of jobs dropped by only 80. Despite this turn of events, the union was operating a rotation policy in 1951 which required each fisherman to lay off every sixth trip. Hence the unionized ports were not suffering from inadequate manpower supplies, although the union was effectively controlling both the number of jobs and the number of fishermen.

Despite this, the union has made little attempt to install systematic hiring practices and procedures. Though the branch agents check often to see that vessels have union-manned crews, union officials do not usually provide men for each trip. In Boston, for example, hiring takes place frequently in a barroom opposite the Fish Pier, or on the adjacent sidewalk. The captain is never sure of his crew until he gets the men aboard ship. Should two or three of the men change their minds about going and not appear at the vessel during the 2 P.M. to 5 P.M. sailing time,[78] there is no ready source to call for replace-

---

[75] Data from the U.S. Tariff Commission.
[76] Data from the U.S. Tariff Commission.
[77] Section on the "profit-sharing lay."
[78] Union contracts specify this period.

ments. As a result, "stragglers" sometimes hold up sailings for a day or more.[79]

PRODUCTION CONTROL

Selective production is as old as the fishing industry itself. With their remuneration on a share basis under the lay system, fishermen have sought always to catch and land only those species for which a market existed. Up to 1818, for example, New England fishermen caught mackerel for bait purposes only;[80] haddock, very abundant during the nineteenth century, were landed in limited quantities only until the twentieth century because the fish was not as well suited to preserving methods as cod, and the market for fresh fish was limited to seacoast areas.[81] In modern times, the redfish was discarded until it became a profitable species after 1935.[82] Pollock, hake, and cusk have yet to gain real favor among consumers. Red hake, for example, remained unutilized until wartime shortages created an unusual demand for fish in 1943 and 1944.[83]

Thus production regulation is an almost natural concomitant of fishery operations. In the modern New England fishery, where nearly all vessels have ship-to-shore telephones and radios to report market conditions, the vessels' captains, with or without instructions from shoreside boat owners, are pretty well equipped to determine what species to catch; what quantities of each species to save and ice down; how soon to return to port. One must bear in mind, however, that these decisions depend to some extent on the "fishermen's luck" of each voyage. For example, the vessel may encounter principally

[79] Under union contracts, vessels may sail short-handed if insufficient numbers require, but both captains and men are somewhat reluctant to do so. Moreover, union contracts do allow nonunion men to sail when full union crews are not available. However, such nonunion personnel are not always present because of the uncertainty of job opportunities; they are not likely to be as competent as union men; the crew may resent their presence aboard ship. For these reasons the captains are often unable or reluctant to hire them.

[80] Cf. Raymond McFarland, *History of New England Fisheries*, page 190.

[81] Cf. H. F. Moore, "The Haddock: One of the Best Salt-Water Fishes," Economic Circular No. 47, U.S. Dept. of Commerce, Bureau of Fisheries, August 18, 1920.

[82] See Chapter III.

[83] Cf. *Fishery Resources of the United States*, Senate Document No. 51, 79th Congress, 1st Session, March, 1945, page 58.

cod, hake, and pollock when the captain is employing all his ingenuity to find large supplies of haddock.

It is against this background that the members of the Atlantic Fishermen's Union have voted catch regulations periodically as their final weapon in the fight to obtain "fair" prices from the dealers. It is important that we stress this, for the fishermen have not adopted restrictions as part of a plot to pick the public's pocketbook. They insist that their incomes suffer through price drops when supplies landed are large;[84] they charge that lower producer prices are not reflected at the consumer level where they might widen the market.

Generally, the union's production controls have taken two forms. They have specified either the type and quantity of fish which the fishermen will land, or they have limited the total catch indirectly through restrictions on the amount of fishing.

Direct controls on output were by far the more common before June, 1947. Whenever the fishermen in a particular port felt that all fish or a particular species were too plentiful, they voted simply to restrict the quantity which could be landed per trip. Exhibit VI displays several such limits adopted by the union between 1941 and 1947. These policies necessitated some dumping of fish at sea; the fishermen could not set their nets to pick up the proper species in exactly desired amounts. On occasion, the fishermen reinforced these limits by stipulating prices below which they would refuse to unload any fish caught. They had to abandon these techniques, however, when the Massachusetts courts intervened to enjoin them in the proceedings set forth in the next section of this chapter.

Undaunted, the fishermen retained their "fair price" concept and turned to indirect methods of restricting the catch. They adopted limits on trip length, where such action seemed desirable. Where it was not, they developed the practice of spending from two days to a week ashore between trips.

Price and production data available for periods following the imposition of some of the restrictions listed in Exhibit VI show that landings nearly always declined. For example, under the limit voted in New Bedford in August, 1942, September landings dropped 800 thousand pounds below the August figure. Under the restriction im-

---

[84] That is, the fishermen view the dealers' demand for fish as inelastic.

EXHIBIT VI

SELECTED CATCH RESTRICTIONS, IMPOSED BY THE MEMBERS OF THE ATLANTIC FISHERMEN'S UNION, ON TYPES OF FISH TO BE LANDED AT BOSTON, GLOUCESTER, OR NEW BEDFORD 1941–1947

Date Imposed	Port	Limit
June 1, 1941[a]	New Bedford	Mackerel limited to a maximum of 40,000 pounds per trip
June 23, 1942[b]	Gloucester	Whiting limited to maximum of 30,000 pounds per trip; no more than 20,000 pounds to be steak whiting; no more than 10,000 pounds to be round whiting
August 20, 1942[c]	New Bedford	Total catch limited to 5,000 pounds per man per boat
April 1, 1945[d]	Boston	Codfish limited to a maximum of 100,000 pounds per trip
January 17, 1947[e]	New Bedford	Yellowtail limited to a maximum of 10,000 pounds per trip
February 7, 1947	Boston	Codfish limited to 5,000 pounds per man per boat Pollock limited to 10% of total catch per trip
April 12, 1947[f]	Boston	Catch per trip limited to 6000 pounds per man Codfish limited to 2000 pounds per man per trip Pollock landings barred
March 19, 1947[g]	Gloucester	Redfish limited to 5000 pounds per man per trip Groundfish limited to 6000 pounds per man per trip

[a] "Cost of Production and Distribution of Fish in New England," *Report of the Federal Trade Commission on Distribution Methods and Costs,* June 30, 1945, Washington, D.C., page 60.

[b] *Gloucester Daily Times,* June 23, 1942.

[c] Telegram, New Bedford Seafood Producers' Association to Charles W. Triggs, Fish Advisory Consultant, OPA, Washington, D.C., August 20, 1942.

[d] Commonwealth of Mass. v. McHugh et al. Suffolk Superior Court, Boston, Mass., June, 1947.

[e] *New Bedford Standard Times,* January 17, 1947.

[f] Commonwealth v. McHugh, *supra.*

[g] *Ibid.*

posed in Boston in February 7, 1947, the pollock catch for that month fell one million pounds from the January level. It is not clear that the union's limits have favored fishermen's incomes. In both the cases cited above, the value of landings available for the fishermen to share declined after initiation of the controls. The fishermen maintain steadfastly, however, that limits are necessary even where reduced catch values may accompany them. Without the union's production regulations, the fishermen argue, sharp price drops would cut landings value far more.

Whether direct or indirect, the union's regulations differ essentially from output decisions which the individual fishing captain might make at sea in the absence of union directives. The union rules usually apply across the board to all vessels in a given port and are set in advance of actual fishing operations. Individual captains' moves, on the other hand, vary from vessel to vessel; they are made in the light of the market and fishing realities of a particular trip.

The effects of restrictive policies cannot be measured with precision. They do, however, raise vessel owners' unit costs. Overhead is an important item in the running of fishing vessels. Policies which reduce their productive use or cut the volume of fish landed must result in higher unit costs of production. Moreover, direct controls on output raise the unit operating costs for which the fishermen are responsible. The trawlers travel hundreds of miles each trip; much of the mileage is to and from the fishing grounds. With direct limits in force, fuel, lubricating oil, and ice, charges per unit are higher.

Indirect controls on output are not much better from the fishermen's standpoint. Trip length regulations — for instance, eight days as a maximum — cut the amount of fishing time against which the costs of sailing to and from the banks may be applied. Also, such regulations are inflexible; they reduce income on those voyages which encounter poor fishing, for they prevent a longer sojourn on the fishing grounds. Finally, tieing up the vessels for several days between trips requires the fishermen to sacrifice some regularity of income.

It is difficult to say whether the union officials, and especially the individual fishermen, have recognized the costs inherent in their production control policies. But even if they have, their suspicion and animosity toward the dealers has remained dominant. Their belief that the dealers "chisel" on prices has confirmed their faith in using limits. This is illustrated clearly by the following passages from a letter on the subject written by the union's secretary-treasurer: [85]

The buyers said they could not pay 3 cents per pound for fish except on a few boats, but that they would buy all the fish that was in the port that day for 2½ cents per pound. The men refused to unload the fish at these prices . . . Therefore, no fish were bought or sold on that day .

On the next day, the buyers bought all the previous day's fish that was in the port at 3 cents per pound, and bought that day's arrivals at 3¼ cents per pound, which to us, at least, proves that it is a case of the buyers trying to chisel on the prices to the fishermen.

When the men saw what had happened in this case, they then agreed among themselves that as the buyers could not handle their fish at a fair price, they would stay in port an extra day (each trip) until the glut was off the market . . .

As for the fish buyers, they have told us that it is their policy to buy the fish as cheap as they can, and we would expect them to sell them as high as they could. So whether they buy them for one cent or three cents from the fishermen, it would not seem to make any difference to the consumer.

The latter paragraph suggests that production limitations have little or no effect on the dealers' competitive position. This is not true. Restricted fish supplies tend to raise the dealers' raw materials costs, to increase their operating costs by lessening the smoothness of work flow, and to step up their unit costs by limiting the volume of output to which overhead might be spread.

There is, however, a germ of truth in the union's charge that whether the dealers buy fish for one cent or three cents, it will make no difference to the consumer. Changes in fish costs at the dock take considerable time to reach the frozen fish consumer because of the factor of cold storage holdings. Millions of pounds of frozen fish are constantly in inventories of main port dealers and wholesalers throughout the nation. Existing stocks have to be worked off to some extent before primary market price changes may filter through the distribution system. The process of primary market price-change transmission is not crystal clear, however, because of the time factor. A future favorable change in demand, for example, may enable some

[85] Excerpts from letter of Patrick J. McHugh to Mr. Lindgren, U.S. Fish and Wildlife Service, Boston, Mass., August 28, 1942.

sector of the distribution stream to pocket better profits because of lower prices paid earlier to the fishermen, or it may save either the primary port dealer or the wholesaler from receiving lower net returns (or losses) because of higher fishermen's prices earlier established. This feature of fish marketing has grown in importance as New England's frozen fish production has expanded constantly during the last quarter century. Gloucester provides the most outstanding illustration. That port markets about 90 per cent of its redfish fillet output in frozen form. The problems of demand, distribution, and profit margins will be discussed in more detail in the next chapter.

It should be pointed out, however, that instances have arisen where the fishermen have had no choice other than to limit their catch. At times during 1944 and 1945 there were more fish landed than could be stored in cold storage facilities in Gloucester.[86] About 100,000 pounds of fish spoiled in New Bedford in the same years for similar reasons.[87]

Apart from income aims, the union has argued that its production controls are favorable to conservation. The serious need for the latter will be explained in the next chapter. While restricted fishing activity must lessen the drain on the resource, it is not necessarily true that the reduced intensity of fishing effort according to the union's policies is very helpful for conservation. The restrictions do not always reduce the catch of scarce species. Fish discarded at sea to implement restrictions usually are thrown back dead because they do not survive the catch process. Finally, the union's limits have not become part of a systematic conservation program.

Union Policy and Public Policy

In 1947 the Massachusetts Superior Court barred the Atlantic Fishermen's Union from employing direct production and marketing controls as weapons in the union's continuing price struggles with the dealers in the principal Massachusetts ports.[88]

---

[86] See the testimony of W. F. Royce, Head, North Atlantic Fishery Investigations, U.S. Fish and Wildlife Service, in *Commonwealth of Massachusetts v. Patrick J. McHugh et al.*, Suffolk Superior Court, Boston, Mass., June 1947, page 1226.

[87] *Ibid.*, page 1228.

[88] Cf. *Commonwealth of Massachusetts v. Patrick J. McHugh et al.*: "Inter-

A set of union acts provided the basis for the long, costly legal struggle initiated by the Massachusetts' Attorney General. First, the union had adopted several catch limits between 1944 and 1947 to bolster dealer-paid fish prices. Second, its members had refused to unload vessels at Boston Fish Pier during the third week of February, 1947 because the prices bid for haddock and cod were less than the winter ceilings set by the Office of Price Administration during the second World War. The latter regulations had expired in May, 1946, but the Boston fishermen had voted on February 7, 1947 to make the wartime ceilings the lowest prices which they would accept. Third, in the same month, the union had allegedly restrained fishing captains from selling to the largest Gloucester firm which had offered the highest bids in the union's selling room. The union insisted that it had a labor dispute with the latter company, because the concern previously deducted more than 5 per cent from the catch weight in paying for a trip purchased in the selling room.[89] The 1947 disputes in Boston and Gloucester touched off the legal "fireworks."

The court condemned all these practices under the Massachusetts antimonopoly statute [90] because they might unduly enhance the price of fish — an essential food within the Commonwealth. Against the union's strenuous objections, the judge ruled that the defendants were subject to the state's jurisidiction. To quote the court, "the defendants' transactions in fish end with the sale and delivery of the fish in Massachusetts to buyers who receive and pay for them here. It is a matter of indifference to the defendants what buyers do with their fish. The defendants in their dealings in fish . . . are not engaged in interstate commerce." [91] Furthermore, the court decreed that it was impossible to find that the fishermen's acts had caused higher fish prices to the Massachusetts consumer but it was sufficient for conviction that such prices might have been artificially raised.

---

locutory Decree Granting a Preliminary Injunction," February 21, 1947; "Findings of Fact and Order for Decrees," Equity No. 58644, July 31, 1947. (Both in Suffolk Superior Court, Boston, Mass.)

[89] See sections on Gloucester and New Bedford in this chapter.

[90] Cf. *Massachusetts General Laws*, Tercentenary Edition, Chapter 93, Paragraph 2.

[91] Cf. J. Broadhurst, "Findings of Fact and Order for Decrees," *Commonwealth of Massachusetts vs. Patrick J. McHugh et al.*, Equity No. 58644, Suffolk Superior Court, Boston, Mass., July 31, 1947, page 24.

The decision ordered the union and its officers "to cease and desist from conspiring, confederating, or combining to limit the production by them of fresh fish for the purpose of having the effect of preventing the sale of fresh fish at its fair market value when offered for sale or preventing a free and fair market in the selling of fresh fish within the Commonwealth." [92] In addition, the decree fined the union $500 for enforcing a limit on pollock after issuance of the original restraining order.[93]

Of the many intriguing facets of this case,[94] two particularly stand out. First, the court refused to concede that special circumstances in the fishing industry — the use of the profit-sharing lay to remunerate fishermen and the direct market contact of the fishermen and fish buyers — made fishermen-dealer controversies a "labor dispute." Second, the state's policy triumph failed to weaken the union's confirmed belief in the wisdom of limiting output to protect the fishermen's earnings.

The first matter played a crucial role in the case. Concession to the union's argument would, in all probability, have exempted the organization from prosecution. The state's law forbids the issuance of injunctions in "labor disputes." [95] The court reasoned that the terms of the lay furnish the fishermen's wages;[96] therefore, the running controversy between the Fishermen's Union and the dealers, which gave rise to the union's restrictive policies, was not a "labor dispute." The union protested in vain that prices are wages under the lay, even where the prices are paid by nonowners because the prices are a payment for the expenditure of labor. The court denied the claim; it admitted that the fishermen's wages, like those of all employees, are derived from the demand for the product on which the fishermen work. But the court insisted that the price conflicts did not constitute a controversy over terms and conditions of employment between union workers and their employers. The judge char-

---

[92] *Ibid.*, page 25.

[93] *Ibid.*, pp. 27 and 28.

[94] For example, the issue of intra- vs. interstate commerce; the question of criteria for determining "fair market value"; and the union's claim that it was only defending itself against alleged collusion of the fish buyers.

[95] Massachusetts has a "baby Norris-LaGuardia Act." See "Findings of Fact and Order for Decrees," page 22.

[96] *Ibid.*, page 22.

acterized the conflict between the fishermen and dealers as "an encounter between the Union and the law of supply and demand . . . a skirmish in the universal contest between seller and buyer . . ."[97]

In effect, the court told the union that it could turn only to the vessel owners, as the direct employers of the fishermen, for satisfaction on wage matters. Put another way, the judge's finding warned the union to act conventionally — to seek as large a share of the "product" price from the vessel owners as the union's bargaining power might warrant, but not to try and enlarge the "product" price to be shared by adopting restrictive policies aimed at the dealers who buy the catch. The latter activities were declared illegitimate even where the buyers are dealer-owners, because, the verdict implied, the latter act as product buyers, not as employers, in bidding for fish. The union's insistence that it sought to squeeze dealer margins, not to force the public to pay more for fish, did not sway the court's judgement.

It might have been better, at least closer to economic realities, if the court had admitted that the profit-sharing lay combined with the primary market structure which brings fish dealers and the union into direct contact does make the union interested directly in price determination. As we have seen, the price-wage relationship is peculiarly intimate under the lay system in New England.[98] Particularly is this so in Boston where most of the trawlers are dealer-owned.[99] The fishermen's restrictive policies were wage-oriented in fact because fish prices and the lay split — as well as the other variables listed on page 59 of this chapter — do converge to produce fishermen's wages in the principal port primary markets. In view of this, dealer-fishermen price conflicts are in large measure wage disagreements; seemingly they are "labor disputes." The union's restrictive policies are extraordinary union policies in an extraordinary industry.[100] The court, however, felt entitled on legal grounds to distin-

---

[97] *Ibid.*, pp. 22 and 23.

[98] See the first section of this chapter.

[99] See Chapter III.

[100] Similar price difficulties have beset West Coast fishermen and dealers and have raised issues of public monopoly policy. The basic difference between the West Coast and New England is that 20 per cent or more of the fishermen on the West Coast either own or lease boats (mostly small vessels); in New England the fishermen clearly are not vessel owners or operators. Consequently, on

guish the dealer-fishermen tiffs as "skirmishes in the universal contest between seller and buyer." This made the union's policies amenable to public regulation. The decision will stand, for the Massachusetts Supreme Court has upheld it and the United States Supreme Court has refused to review the case.[101]

The court's legal reasoning provided the state with a "victory"; it did not convince the union of the unwisdom of the policies involved. The union spent very heavily to fight an unsuccessful appeal. Meanwhile, it adopted indirect methods which limited trip lengths and sailings, to control production where the fishermen thought it was necessary. Since these restrictions are voluntary limitations upon the amount of work the fishermen are willing to do, they would seem to be legal.[102] Should these be challenged successfully by public authority, leaders of the union have announced, then the organization will find other ways to "protect" the men's earnings from dealer "manipulation." [103]

The "other ways" might very well include an industry-wide strike for a straight wage-payment system.[104] Such a turn of events would meet with little in the way of a happy welcome in the industry. It is no accident that the lay, in a variety of forms, has persisted as the predominant method of payment. The owners would not be sanguine about the shift; many of them literally are "wedded" to a lay system

---

the West Coast, an additional difficulty is whether the fishermen are "independent businessmen" or employees. See Roger L. Randall, "Labor Agreements in the West Coast Fishing Industry: Restraint of Trade or Basis of Industrial Stability?" the *Industrial and Labor Relations Review* (Cornell Univ. Press), vol. 3, no. 4, July, 1950, pp. 514–541. See also Pinsky, Goldner, and Eden, *The Fisheries of California, the Industry, the Men, Their Union*, California CIO Council, Research Department, San Francisco, Calif., January, 1947.

[101] Cf. *Commonwealth of Massachusetts v. McHugh et al.*, 93 N.E. 2d 751, July 7, 1950, *North Eastern Reporter*, Second Series (St. Paul, Minn.: West Publishing Co., September 13, 1950), pp. 751–768. The Massachusetts Supreme Court made only the distinction that nothing in the decree should be construed to prevent individual captains and crews, acting independently of other captains and crews, from selling their particular catches at the highest prices obtainable.

[102] See, however, the current New Bedford indictment.

[103] This was the opinion expressed by Patrick McHugh, Secretary-treasurer, in a strong statement before the Joint Senate-House Committee to Investigate the Massachusetts Fishing Industry, State House, Boston, Mass., September 29, 1947.

[104] In fact, McHugh stated flatly, in an interview with the author on September 19, 1952, that this would happen.

— particularly the smaller vessel operators — for a few bad trips with the overhead of wages to meet would break them, as even the union admits. A strike to secure straight wages might well be catastrophic, for the union leaders have made it clear that the men would insist on average earnings somewhat approximating existing earnings under the lay. The fishermen, of course, would be delighted to work for straight wages under the aforementioned condition.

It might be argued that a straight wage system would benefit the industry by wiping out "marginal" operators. There is some truth to such a statement. To the extent that such a development cut output, however, it would be like a sharp union restriction, applied instantaneously. In terms of satisfying the principal economic groups, such a move would rate a very low score. Dealers and shoreworkers would scream because of any diminution in production; bitterness would be rife among the vessel owners. As for the Union, we can judge the results best perhaps by recognizing that for years the organization has had the power within its grasp to shut down a fairly good number of vessels simply by refusing to allow the members to sail on craft that did not measure up fully in terms of seaworthiness. But the union is interested in jobs, and hence on the latter point has tried to keep safety demands within the capacity of most vessel owners to meet them — in other words to achieve the maximum in lifesaving equipment, and so on, consistent with the owners' ability to provide such devices.[105]

Perhaps the most important drawback of a straight wage proposal would be its impact on fisherman incentive and effort. Even McHugh admits that the men's hearts might not be in a task presented at midnight under relatively stormy conditions on the banks if the men were on "wages." Yet, as happens often, such conditions might be prevailing when for the first time during a trip the fish are beginning to "run good."

To return to the main thread of our story, it seems clear that antitrust action offers no "cure" for the problems of the fishing industry. Indeed, when one adds the bitterness, vituperation, aggravation of old "grudges," and stimulation of sour memories which were par-

---

[105] In addition, the union has attempted to secure federal inspection for fishing vessels, as we noted earlier in this chapter.

ticularly associated with the 1947 trial in Massachusetts, it would seem that state intervention may well have delayed a higher level of accommodation to the interests of the parties. Thus monopoly prosecution stands as ill-fitted for the task of inducing a determined trade union to shift its thinking to more progressive terms. Positive governmental action is needed to help eliminate the causes of short-sighted restrictionism. Governmental aid, however, can be effective only with a reconciliation of differences within the industry itself.

Chapter VII will examine the possibilities for such a reversal of form in the major ports, but first we turn to an analysis of the fish scarcity, marketing, and foreign competition problems which harass New England fishing interests.

CHAPTER VI

SUPPLY AND MARKET PROBLEMS

WHILE INTERNAL CONFLICT plagues the principal New England fishing ports, three formidable problems threaten their very capacity to
survive as major fishing centers. First, declining abundance has
raised the cost of exploiting the chief species in the New England
catch. Second, marketing difficulties have limited the sales income
of the industry. Third, growing imports jeopardize the sales of New
England's packaged fish in the United States market.

What are the distinctive features of these pressing problems? What
reactions have they produced in the industry? What steps do they
make imperative?

DECLINE IN FISH POPULATION

As Chapter I has indicated briefly, three of the most important
food fish in the New England catch have grown scarce in recent
years.[1] They are haddock, Boston's specialty and the area's most
valuable groundfish;[2] redfish,[3] the mainstay of the frozen-fish industry at Gloucester and Portland; and yellowtail flounder, a highly
valuable specialty of New Bedford.

United States Fish and Wildlife Service biologists, noting the depletion of the older North Sea fishing grounds under the impact of

[1] See pp. 4 and 5. Other species, particularly whiting, the blackback flounder,
and mackerel pose supply problems for the industry. The situation for the first
two is somewhat similar to that for the other groundfish, but little has been done
about it. The mackerel, a seasonal species, is subject to wide and as yet unexplained variations in abundance. In the text we shall confine discussion to the
"key" species, where an effort is being made to solve the problem.

[2] "Groundfish" are found in all seasons. They include principally haddock,
cod, redfish, hake, pollock, whiting, flounder, and cusk.

[3] "Redfish" generally are marketed in frozen filleted form under the name
"ocean perch."

heavy trawling operations,[4] initiated studies of the New England resource decline. While they have not determined definitely the reasons
for the scarcity of key species, much of the evidence which they have
uncovered to date indicates that overfishing and wasteful fishing
practices are contributing factors.

HADDOCK

The New England haddock fishery has been a source of grave concern to federal fishery biologists since about 1930. The haddock catch
per day on a standard group of Boston otter trawlers during the
"haddock year" 1930 [5] dropped nearly 50 per cent from 1929 levels
as Exhibit VII indicates. Moreover, the Exhibit shows that the 1930
performance of the haddock fishery was only about one-fourth as
effective as in 1927, the peak year.

By 1935, U.S. Bureau of Fisheries [6] biologists suggested two principal reasons for the scarcity of marketable haddock on Georges
Bank, the nearest and most prolific New England fishing ground.
First, haddock spawning between 1925 and 1928 had failed to add
appreciably to the haddock stock. Second, the fishing fleet had removed haddock from the bank so rapidly that the annual commercial catch had more than tripled in ten years. This quickly undermined haddock abundance.[7]

The first condition, spawning failure, appeared to be beyond control. No practical method of influencing offshore spawning has yet
appeared. The second factor offered more opportunity for action.
Bureau of Fisheries investigations revealed that millions of pounds
of less than marketable size baby haddock were being captured and
destroyed by trawler fishermen. These small, immature haddock were
swept from the sea and thrown back dead.

After extensive tests, the biologists recommended that the indus-

---

[4] Cf. E. S. Russell, *The Overfishing Problem* (London: Cambridge Univ.
Press, Bentley House, 1942).

[5] The "haddock year" runs from February to, and including, January of the
following year.

[6] The Bureau of Fisheries was the predecessor of the Fish and Wildlife
Service.

[7] Cf. W. C. Herrington, "Modifications in Gear to Curtail the Destruction
of Undersized Fish in Otter Trawling," *Investigational Report No. 24*, U.S. Dept.
of Commerce, Bureau of Fisheries, 1935, page 3.

try adopt a larger mesh net to allow baby haddock to escape capture until they reached marketable size. Experiments with commercial fishing gear disclosed that the smaller mesh employed in the industry caught about five times as many small haddock as the recommended larger mesh.[8]

EXHIBIT VII

TOTAL HADDOCK CATCH PER DAY FOR STANDARD GROUP OF BOSTON OTTER TRAWLERS ON GEORGES BANK, 1917–1948

(*pounds*)

Year	Catch
1917	25,825
1918	32,950
1919	34,975
1920	36,575
1921	32,475
1922	24,475
1923	18,375
1924	23,150
1925	32,250
1926	41,275
1927	43,790
1928	34,520
1929	22,402
1930	11,545
1931	8,880
1932	11,572
1933	9,708
1934	10,308
1935	12,275
1936	13,500
1937	11,650
1938	11,733
1939	13,040
1940	12,836
1941	16,615
1942	18,682
1943	18,343
1944	16,973
1945	16,000
1946	14,264
1947	12,800
1948	9,702

Source: U.S. Fish and Wildlife Service Biological Laboratory, Woods Hole, Massachusetts.

[8] Cf. W. C. Herrington, "Decline in Haddock Abundance on Georges Bank and a Practical Remedy," *Fishery Circular No. 23*, U.S. Dept. of Commerce, Bureau of Fisheries, July, 1936, page 18.

The industry spurned the mesh recommendation. Vessel owners could not agree unanimously to make the change. Each feared that the other would secretly use the smaller mesh and gain a competitive advantage. Also, all vessel operators claimed that the larger mesh would give an advantage to unregulated foreign fishing interests operating on New England offshore banks. The domestic boat owners saw no profit in conserving haddock for the benefit of foreign fishermen. With little success, the Bureau of Fisheries countered vessel owners' objections by insisting that larger mesh nets would actually catch more large haddock than do the smaller type, probably because of a better draft of water through the larger mesh. The agency argued further that dragging the larger mesh would subject the vessel to less strain and reduce mechanical wear.[9]

Instead of heeding the suggestion, the industry stepped up landings from Canadian waters and concentrated upon scrod [10] from the New England banks. Unfortunately, neither policy helped the resource situation. Haddock became less plentiful on the nearer Canadian banks. Between 1931 and 1941, the catch per day of large haddock advanced only about 20 per cent from the low point hit in the former year; the catch per day of scrod increased six-fold. Scrod accounted for only 12 per cent of Boston haddock landings in 1931; in 1941 it accounted for about 40 per cent. The destruction of baby haddock on Georges Bank through heavy scrod fishing cut the haddock population on that fishing ground to one-third its size in earlier years, according to the U.S. Fisheries Laboratory at Woods Hole.[11] The same authority estimated that between March and October, 1947, about 15 million baby haddock averaging less than one pound in weight were killed and discarded on Georges Bank by the New England otter trawler fleet. If these fish had not been caught until the next spring, they would have increased the catch by at least 20 million pounds. Assuming 1947 prices, this would have brought an additional $1.5 million to the industry's fishermen and vessel owners.[12]

---

[9] *Ibid.*

[10] In the industry haddock weighing from 1½ to 2½ pounds, inclusive, are called scrod.

[11] Cf. H. A. Schuck, "Current Haddock Situation on Georges Bank," *Commercial Fisheries Review* (Washington, D.C.: U.S. Fish and Wildlife Service, October, 1948), page 1.

[12] Cf. H. A. Schuck, "Protecting Baby Scrod Raises Production," U.S. Fish and Wildlife Service, Washington, D.C., 1947.

It is difficult to estimate how costly the scarcity has been for Boston, New England's haddock center. The effect of higher costs of sailing greater distances to Canadian banks has varied annually with the fluctuations in scrod abundance in New England waters. At any rate, landings of haddock from Nova Scotian waters increased from 9.6 million pounds in 1930 to a peak of 64.2 million pounds in 1935. For the decade of the thirties they averaged 37.3 million pounds, well over three times the level of 1930.[13] In 1934, a U.S. Bureau of Fisheries spokesman estimated that during 1934 alone the longer trips to Nova Scotia banks added more than 2600 days to the running time of Boston trawlers at costs he estimated to be about $250 per day.[14] During the second World War, when about two-thirds of the Boston fleet ceased fishing to enter war service, landings originated mainly from New England waters.[15] During this period, with the decreased fishing effort, output per day of fishing rose sharply on the New England banks, but it dropped off in the postwar years as the fleet regained somewhat its prewar strength. In the postwar period, haddock landings from Canadian waters have averaged just over 33 million pounds annually and once again the more costly, longer trips have returned.[16]

Boston suffers in another way, too, from this situation. Longer trips mean decreased quality of fish landed. The increased volume of scrod in Boston's landings tends to tie that port's fortunes partly to the fortuitous success or failure of the spawning process. In a year of poor scrod abundance the industry faces no choice other than to expand trips to more distant banks. There is another somber aspect to the overwhelming importance of scrod in Boston's landings. Early in 1952 scientists of the North Atlantic Fisheries Investigation of the U.S. Fish and Wildlife Service at Woods Hole, Massachusetts, reported that the haddock fisheries on Georges Bank, according to their most recent analysis, had stabilized — that is, that there would

[13] Data are from the U.S. Fish and Wildlife Service. They cover Boston, Gloucester, and Portland, but the lion's share of the catch was landed in Boston.

[14] Cf. W. C. Herrington, "Decline in Haddock Abundance on Georges Bank and a Practical Remedy," *Fishery Circular No. 23*, U.S. Department of Commerce, Bureau of Fisheries, July, 1936, page 9.

[15] Fishermen stayed in New England waters because of the U-boat menace as well as because of the improved yield.

[16] Data are from the Fish and Wildlife Service and cover the principal Massachusetts ports. Again the principal part of the total landed at Boston.

not be a further decline in production with the present intensity of fishing. While this sounded like good news, nevertheless, it meant also that (a) increased fishing effort on Georges Bank might very likely cut output per man day of effort quite considerably and (b) under present conditions the industry must look forward to capturing a sizeable percentage of its haddock landings in Canadian waters. For the year 1951 not quite 25 per cent came from that source. Moreover, in view of the fact that scrod accounted for about 55 per cent of haddock landings from New England grounds in 1950 and for the first time in history exceeded the landings of large haddock, the stability reported by the Woods Hole scientists will nevertheless probably yield fairly large annual variations in productivity.

Secondly, scrod is somewhat less desirable than large haddock, for it is more difficult to handle and process, and generally yields not quite so good a product.

U.S. Fish and Wildlife Service scientists have continued to recommend adoption of larger mesh nets in the haddock fishery. They now have clear evidence that haddock grow most rapidly during their first three years of life. On Georges Bank, the average haddock quadruples its weight between its first and second years of life; it doubles its weight between the second and third years.[17] This early and rapid rate of growth puts a premium on avoiding the capture of immature fish. Moreover, Fish and Wildlife biologists emphasize that larger mesh nets have other advantages: they would cost less; they would handle more easily because they would pick up less trash from the ocean floor; they would mend more easily, for they have fewer knots; they would allow increased towing speeds, for they are lighter. Up to 1950 these arguments failed to move the industry's vessel owners. Fishermen's union officials were sympathetic and supported generally the biologists' stand. The union did not try, however, to make the conservation proposal a matter for collective bargaining.

Fish and Wildlife scientists proffered two other conservation suggestions as early as 1936.[18] The first was a minimum legal length and weight specification for haddock landed, to discourage the taking of

[17] Cf. H. A. Schuck, "Current Haddock Situation on Georges Bank," *Commercial Fisheries Review* (Washington, D.C.: U.S. Fish and Wildlife Service, October, 1948), page 5.

[18] Cf. W. C. Herrington, "Decline in Haddock Abundance on Georges Bank and a Practical Remedy," pp. 20–22.

immature fish. The second was the negotiation of an international conservation treaty to arrange coördinated exploitation of a sustained yield from the North Atlantic fishery resource by the several nations drawing upon it. All three proposals have had the support of the Atlantic States Marine Fisheries Commission. The latter organization is composed of industry and public delegates from the several North Atlantic states. It was set up in 1942 under a congressionally-approved compact [19] to coördinate the fisheries policies of the member states. Contributions from the latter support the commission's operations.

Action on the treaty recommendation started in 1949. The several countries [20] whose fishermen depend upon the North Atlantic fishery resource negotiated the International Convention for the Northwest Atlantic Fisheries.[21] The signing of the treaty document was a considerable achievement. The United States Tariff Commission has pointed out the abiding difficulties of reaching international fisheries agreements, particularly in our northeastern fisheries, which have figured so prominently in international relations.[22]

The convention establishes an International Commission to deal with conservation problems. Each contracting government has one vote in the body, and not more than three commissioners. Further, the convention divides the Northwest Atlantic into five principal sub-

[19] Cf. Public Law 539, 77th Congress, 2nd Session (56 Stat. 267, 1942). The Commission has taken a very active part in industry affairs, in supporting moves really to undercut such problems as that of fish scarcity. Its stimulating meetings have provided much opportunity for digesting and developing constructive thought on problems of all sectors of the east coast fishing industry.

[20] The countries were the United States, Canada, Denmark, France, Iceland, Italy, Newfoundland (since April, 1949, a Canadian Province by vote of the Newfoundland citizenry), Norway, Portugal, Spain, and the United Kingdom of Great Britain and Northern Ireland.

[21] Cf. "International Northwest Atlantic Fisheries Conference," DOC/ 1–51 (Washington, D.C.: U.S. Dept. of State, Jan. 26, 1949), and *The Fisheries Conventions*, Hearings before the Subcommittee of the Committee on Foreign Relations, U.S. Senate, 81st Congress, 1st Session, pp. 5–21. All information on the convention is from these sources, unless noted otherwise.

[22] Cf. U.S. Tariff Commission, *Treaties Affecting the Northeastern Fisheries*, Report No. 152, Second Series (Washington, D.C.: U.S. Government Printing Office, 1944). This document presents the substance of the various treaties, agreements, and awards which had important effects on the fisheries of American vessels in northeastern waters, and discusses the effects on the fishing industry of the treaties and other documents.

areas; each is to have a special panel of commissioners drawn from the members of the over-all commission. Care is taken to insure that countries substantially dependent upon the fisheries in a given subarea shall be adequately represented on that area's panel. The commission as a whole will coördinate the work of the panels and give preliminary consideration to panel recommendations. The commission will be responsible also for scientific investigations.

The convention does not empower either the commission or the special area panels to take direct regulatory action. The panels may originate conservation proposals; the latter may not become effective until four months after notifications of acceptance thereof have been deposited with the United States government by all the contracting governments participating in the panel or panels for the subarea or subareas to which the proposals apply.[23]

The chief merit of the convention is that it provides machinery for international coöperation in the scientific investigation and development of the Northwest Atlantic fishery resource. The attempt of the United States, in its initial draft for consideration of the conference delegates,[24] to provide specific machinery for enforcement of the convention's terms, did not work out. The convention binds the contracting governments merely to take such action as may be necessary to make effective its provisions and to report statements of action taken to the commission.

The Atlantic States Marine Fisheries Commission, the vessel owners' associations in the principal New England ports, and the Atlantic Fishermen's Union nearly scuttled implementation of the convention by the United States. They agreed with the convention in principle, but they clashed sharply on enabling legislation, especially on the matter of selection of the three United States commissioners. Each thought one of the three commissioners should come from its ranks. In turn, the union opposed strongly any representation from the Atlantic States Marine Fisheries Commission.[25] The bill offered

---

[23] The commission itself may initiate proposals affecting the convention area as a whole after consultation with all the panels.

[24] The United States convened the International Northwest Atlantic Fisheries Conference; W. M. Chapman, Special Assistant to the Under Secretary of State, acted as Chairman.

[25] Cf. *Northwest Atlantic Fisheries Convention*, Hearings on S. 2801, Sub-

in the U.S. Senate called for one representative from the U.S. Fish and Wildlife Service; one commissioner drawn from the official fishery agencies of the states whose vessels fish in the convention area; and a third drawn from the public at large. The Senate Committee on Interstate and Foreign Commerce resolved the log jam by stipulating that the president of the United States make the selections at his pleasure. The positions taken by the vessel owners and the union were an attempt to insure that they have an adequate voice in determining any regulations under which they would have to operate. The revised legislation provided that the fishermen and vessel owners have fair representation on a five- to twenty-person board of advisors to the three commissioners.

The convention entered into force in July of 1950, after deposit with the United States of ratifications by four signatory nations.[26] In January, 1951, the President appointed the three U.S. commissioners under the treaty, among whom was Francis W. Sargent, Director of Marine Fisheries in Massachusetts and a person rather close to the problems of the New England groundfish industry.[27] At their initial meeting on March 6 and 7, the commissioners named fourteen advisors representing all phases of the North Atlantic coastal industry.

As we shall see in Chapter VIII, developments between March 20, 1951, when the commissioners held their first meeting with the advisory group, and the end of 1952 augur well both for the future of the convention and the New England fishing industry. Moreover, the advance in technical knowledge of the resource should be more rapid in the next several years than perhaps ever before, for the convention will concentrate upon the collection, collation, and dissemination of scientific information.

---

committee of the Committee on Interstate and Foreign Commerce, U.S. Senate, 81st Congress, 2nd Session, April 4 and 5, 1950; also *Northwest Atlantic Fisheries Act*, Report No. 1830, Committee on Interstate and Foreign Commerce, U.S. Senate (Senate Calendar No. 1836), 81st Congress, 2nd Session, June 14, 1950.

[26] The four nations were the United States, Canada, Iceland, and the United Kingdom. We should note that the U.S. Act giving effect to the treaty contained enforcement penalties. See Public Law 845, 81st Congress, 2nd Session, 1950.

[27] The other commissioners appointed were Hilary J. Deason, Chief of the Office of Foreign Activities, U.S. Fish and Wildlife Service, and Bernhard Knollenberg, author and lawyer from Chester, Connecticut.

REDFISH

Scarcity of redfish in New England and adjacent waters is a comparatively recent phenomenon. Intensive exploitation of the resource did not start until after 1935; the heaviest drain has been concentrated in the last ten years. The industry has scored impressive production records. The tremendous expansion in the redfish catch could not have occurred, however, without the help of longer trips to northerly Canadian banks. This is demonstrated clearly in Table 21.[28] As

TABLE 21

REDFISH LANDINGS AT THE PRINCIPAL NEW ENGLAND PORTS, BY FISHING GROUNDS
WHERE CAUGHT, 1935–1951

(millions of pounds)

Year	Off Canada	Off United States
1935	0.57	16.54
1936	15.63	50.97
1937	26.26	32.06
1938	19.47	45.23
1939	21.60	56.01
1940	25.72	58.06
1941	20.71	118.65
1942	4.39[a]	110.76
1943	7.19[a]	94.14
1944	7.80[a]	96.28
1945	41.98	72.67
1946	75.02	82.95
1947[b]	42.26	64.34
1948[b]	112.58	76.36
1949[b]	129.50	51.98
1950[b]	81.37	46.82
1951	119.42[c]	42.31

Source: "New England Landings by Gear and Area at Certain New England Ports," 1935–1946 (Washington, D.C.: U.S. Fish and Wildlife Service).

[a] Low figures accounted for by the second World War. Submarine threat cut sailings to Canadian banks.

[b] Figures for these years for Massachusetts ports only.

[c] An additional 16.67 million pounds were landed from the Newfoundland Grade Bank region.

the reserve of fish had been removed from local grounds, the New England fleet turned to the more distant Nova Scotian banks. By

[28] See also A. Perlmutter, "The Future of the Rosefishery," *Gloucester Master Mariners' Association Yearbook,* 1947, pp. 7–9.

1950 it appeared that the limit of expansion had been reached. A decline in the production of this resource seemed inevitable.[29]

The critical factor in the dismal outlook for redfish supplies is their slow rate of growth.[30] Biologists estimate that these fish require about ten years to reach sexual maturity. The indications are that redfish have been captured at a faster rate than they could reproduce and grow to saleable size. In 1949 the Gulf of Maine, a New England bank, no longer provided full trips of redfish. A few years earlier the bank had been the principal source of supplies. Moreover, vessels fishing on that bank reported the fish found there to be very small.[31] When the more distant banks have been fully exploited and total production commences to decline, it will take considerable time for this fishery to recover. The situation is potentially more dangerous than that in the haddock fishery.

Redfish vessels encounter cost difficulties similar to those cited for Boston haddock vessels, as they push further afield in search of new stocks.[32] Unfortunately, research on the redfish scarcity has not proceeded even as far as the study of the haddock shortage. The U.S. Fish and Wildlife Service initiated investigations of the resource in 1944. They were interrupted in 1949, and then resumed in 1950.[33] There are no official recommendations for alleviating the redfish decline. Thorough analysis of the problem may have to await action under the International Convention discussed in the previous section.

YELLOWTAIL FLOUNDER

Fishermen have sought yellowtail flounder intensively only since the opening of the New Bedford market and the discovery that the fish makes an excellent filleted product.[34] Yellowtail landings at New

---

[29] Cf. "Limit of Expansion for East Coast Rosefishery Reached," *Fishery Products Report No. 185*, U.S. Fish and Wildlife Service, Boston, Mass., September 22, 1950.

[30] Cf. A. Perlmutter and G. M. Clarke, "Age and Growth of Immature Rosefish in the Gulf of Maine and Off Western Nova Scotia," *Fishery Bulletin 45* (Washington, D.C.: U.S. Fish and Wildlife Service, 1949).

[31] Cf. "New England Rosefishery," *Fishery Products Report 218*, U.S. Fish and Wildlife Service, Boston, Mass., November 9, 1949.

[32] See section on haddock, *supra*.

[33] Cf. "Rosefish Investigations Reactivated," *Fishery Products 47*, U.S. Fish and Wildlife Service, Boston, Mass., March 9, 1950.

[34] See Chapter III.

Bedford reached a peak of nearly 37 million pounds in 1942 and then dropped to lower levels from which they have never recovered fully. Only more intensive fishing on southern New England grounds and the discovery of some additional stocks on Georges Bank kept production from further declines after 1944. The average daily catch of yellowtail by small trawlers on the southern New England banks fell from 9600 pounds in 1942 to about 4800 pounds in 1948.[35] Landings from Georges Bank increased from 10 per cent of New Bedford's yellowtail catch in 1946 to 62 per cent during the first eight months of 1949.[36]

United States Fish and Wildlife biologists who were investigating the earlier decline in abundance of the blackback flounder [37] turned their attention to the yellowtail early in the second World War. By 1949 they had discovered that there were three basic stocks of the species; it matures in about three years and is generally unpredictable. The biologists recommended early in 1950 that a minimum size limit of eleven inches on landings of yellowtail would help cushion the impact of intensive fishing on the resource.[38] Yellowtail, like haddock, seem to grow most rapidly in their first three years of life; saving immature fish from capture would add greatly to the productivity of the fishery. The scientists are pressing complete study of the species, but they make no promise of early results. Up to 1952 the industry showed little disposition to try out the scientists' preliminary recommendations.

The industry is beginning to realize, however, that unrestricted fishing for a key species will bring it to grief. Wasteful practices should be abandoned immediately, or at least, recommendations to reduce waste should be put into practice quickly on a trial basis. The industry needs to encourage and coöperate with the accumulation

---

[35] Data from Raymond Buller, Aquatic Biologist, U.S. Fish and Wildlife Service, Woods Hole, Mass.

[36] Data from the Market New Office, U.S. Fish and Wildlife Service, New Bedford, Mass.

[37] Blackback flounders were the more popular variety before the expansion of yellowtail marketing. The decline in the blackback fishery led the industry to discover the merits of the yellowtail.

[38] Cf. W. F. Royce and H. A. Schuck, "Recommendations for Minimum Size Limits on Certain Fishes" (Mimeographed), U.S. Fish and Wildlife Service, Woods Hole, Mass., Feb. 13, 1950, page 4.

of knowledge vital for the setting of regulations designed to promote a high, sustained yield from species which tend to become scarce. Research to improve the efficiency of vessels and fishing methods and systematic investigation to discover new stocks of popular species would pay substantial dividends. Broadening the markets for lesser known underexploited species would help relieve the pressure on haddock, redfish, and yellowtail resources.[39] These steps, of course, are much more easily advocated than accomplished. How might they be initiated? Chapter VII will suggest the avenues.

Marketing Difficulties

New England fishing interests must surmount formidable barriers to maximize the sales value of their fresh and frozen fish in American markets. The competition of other protein foods, principally meat, and the competition of imported fish limit the sales and prices of New England fish.

THE DEMAND PROBLEM

Fish have been called the "bottom rung of the economic ladder of meat." In June, 1947, the National Fisheries Institute reported that average weekly expenditures in the United States for meat and poultry were \$2.36; for fish they were only 20 cents.[40] Moreover, annual per capita consumption of other protein foods — meat, cheese, and poultry — in 1952 was about 187 pounds in contrast to 6.2 pounds per person for fresh and frozen fish.[41] Over-all fish consumption of 11.3 pounds per capita in 1952 made the United States an intermediate fish-consumption nation. We stood far below such industrial nations as Belgium, Germany, and the United Kingdom, which ranked as high consumption areas with per capita consumption ranging from 22 to 43.8 pounds.[42] The overwhelming preference of

[39] Haddock conservation under the International Treaty (*supra*) actually started in 1953. See Postscript to this book.

[40] Cf. "N.F.I. Flashes," National Fisheries Institute, Washington, D.C., June 28, 1947.

[41] Cf. *The National Food Situation*, U.S. Department of Agriculture, Bureau of Agricultural Economics, Washington, D.C., April–June, 1953, page 4.

[42] Cf. "Computing Human Consumption of Fish," *Fisheries Bulletin*, United Nations Food and Agriculture Organization (Quoted in *Fishery Products Re-*

consumers for meat severely restricts the total demand for fish in the United States. At the same time, the presence of so many close substitutes makes the market demand for fish quite elastic.[43]

The income of the fishing industry tends to vary in close conjunction with ups and downs in United States national income over longer periods. An unpublished study by the U.S. Fish and Wildlife Service indicates that the gross income of commercial fishermen varied rather uniformly with changes in national income between 1930 and 1945.[44] Between 1929 and 1932, fisheries income dropped 71.2 per cent while the national income declined about 52 per cent; between 1932 and the period 1935–1939, fish income rose 93.5 per cent while the national income advanced by about 60 per cent.[45] These data indicate that the fishing industry has a real stake in maintenance of sustained, high level national income.

In the past, institutional factors have tied fish demand rather strongly to one day a week and the later winter period, since religious preference and custom have favored fish consumption on Fridays and during the Lenten season.[46] Studies conducted by the U.S. Fish and Wildlife Service during 1939 in 56 large cities in the eastern third of the United States revealed that about half the fish sales of all retail stores surveyed were made on Friday, and nearly one-fifth on Thursday.[47] The figures for Boston, where per capita annual con-

---

port 89, U.S. Fish and Wildlife Service, Boston, Mass., May 6, 1949). The U.S. per capita figure for 1948 was 11.1 pounds, the same as the 1935–1939 average.

[43] Demand is elastic when relatively small changes in price bring about relatively large changes in the amount purchased.

[44] The coefficient of correlation was .99. I am indebted to Dr. Richard Kahn, Chief Economist for the Defense Fisheries Administration and formerly with the U.S. Fish and Wildlife Service for the opportunity to examine the study. The data covered the years 1930–1945.

[45] Cf. E. J. Burtt, Jr., *An Economic Appraisal of the New England Fishing Industry*, unpublished doctoral dissertation, Duke University, 1950. Data are for National Income by Industrial Origin, from the U.S. Department of Commerce.

[46] There are about 26 million Catholics in the United States whose religion forbids them to eat meat on Fridays and on certain days during the Lenten season. See *Official Catholic Dircetory* (N.Y.: P. J. Kennedy, 1948).

[47] Cf. Ralph Russell and K. O. Burr, "Retailing Fresh Fish in 56 Cities in the Eastern Third of the United States," *Fishery Market News*, May, 1942. A similar investigation in five cities of the upper Ohio Valley yielded similar results.

sumption far exceeded the national average, revealed an even greater concentration of retail fish sales on Thursday and Friday.[48]

Between 1939 and October of 1951, the one-day-a-week characteristic of fish demand diminished considerably in importance, though it did not cease to operate as a factor in fish marketing. Using the questionnaire method, Fish and Wildlife officials found that nationally only about 34 per cent of those interviewed served fish particularly on Friday; about 58 per cent of the housewives indicated that they do not have any particular day on which fish is served.[49] The change in consumption pattern was substantially less, however, in the high-consumption northeastern United States region. In October, 1951, in that area about 50 per cent of those interviewed chose to serve fish particularly on Friday.[50]

In the past, too, fish sales in Lent have overshadowed prominently those during the rest of the year. Cold storage holdings of fish drop sharply during the first four months of each year, when the Lenten period occurs, as the industry moves heavy stocks to consumers.[51] Lent acts also as a leverage on fish prices. A study of quarterly average price and production movements for cod and haddock during each year from 1935 to 1941 disclosed that "ex-vessel" prices [52] were consistently higher in the first quarter than would be indicated by the associated production.[53] In other quarters of each year, prices and production tended to move in opposite directions.

In the period after the second World War, similar indicators suggest that the fairly heavy concentration of demand in Lent has diminished. Swings in cold storage holdings, for example, have become somewhat less pronounced so that for the last few years they prob-

---

[48] Idem., "Fish Retailing in Boston, Massachusetts," *Fishery Market News*, August, 1942.

[49] Cf. W. H. Stolting and M. J. Garfield, "Fish and Shellfish Preferences of Household Consumers — 1951," part I, National Summary, and part II, Regional Summary, *Fishery Leaflets 407 and 408*, U.S. Fish and Wildlife Service, May and September, 1952.

[50] *Ibid.*

[51] For the pattern of freezings and cold storage movements, see *Frozen Fish Reports*, U.S. Fish and Wildlife Service, Washington, D.C.

[52] "Ex-vessel" prices are those paid to fishermen and vessel owners.

[53] Cf. W. C. Herrington, "Imported Fish: A Major New England Problem," *Commercial Fisheries Review* (Washington, D.C.: U.S. Fish and Wildlife Service, February, 1946), pp. 1–16.

ably reflect as much the industry's adjustment to production patterns as they do the influence of Lenten consumption.[54] The household interview study previously cited [55] found that consumers currently have no season in which they serve fish more often than during the other seasons of the year.

The evidence suggests, then, that the principal marketing problem of the industry is to increase over-all demand, and to a lesser extent to smooth out the weekly fish consumption cycle at a high level. It must somehow by public education make fresh and frozen fish a matter of everyday consumption.[56] In addition, the industry must encourage the sale of fish at low prices in relation to those for competing foods. In the peacetime period from 1930 to 1941, the consumption of New England's cod and haddock was highest at those times when the prices of those species were lowest in relation to the prices of other competing foods.[57]

The fishing industry must convince the public of the desirability of eating fish and publicize the varied ways in which it may be served. Tests have demonstrated that fish is equal to meat in nutrition; [58] as a marine product, fish is high in mineral content and is an excellent source of vitamins.[59] Nevertheless, housewives seem traditionally to have classified fish as an inferior food to be served only when they

[54] Cf. *Frozen Fish — 1950*, U.S. Fish and Wildlife Service.

[55] Cf. W. H. Stolting and M. J. Garfield, *Fishery Leaflets 407 and 408*, 1952.

[56] According to the noted economist, Roger W. Babson, Chairman of the Board of the Atlantic Coast Fisheries Co., "Too many fish dealers depend upon the Archbishops of the Catholic Church to sell their fish." He reminds the dealers that the hierarchy might change its mind and allow Catholics to eat meat on Fridays! See R. W. Babson, "What Lies Ahead for the Fishing Industry of the U.S.," the *Fishing Gazette* (Annual Review Number, New York, 1947).

[57] Cf. W. C. Herrington, "Imported Fish: A Major New England Problem," *Commercial Fisheries Review* (Washington, D.C.: U.S. Fish and Wildlife Service, February, 1946), page 16. Herrington used primary market price data because of the unavailability of satisfactory retail price information.

[58] Cf. S. J. Wilson, "The Effect of a Seafood Diet on the Red Cell Count, Hemoglobin Value, and Hematocrit of Human Blood," *Fishery Leaflet 334* (Washington, D.C.: U.S. Fish and Wildlife Service, 1949). The article reports results of experiments at the Service Technological Laboratory, College Park, Maryland.

[59] Cf. J. R. Manning, "Look to the Sea for Your Diet," *The Fish and Oyster Reporter* (Tampa, Florida, February, 1933) page 11. Mr. Manning was in charge of technological investigations for the U.S. Bureau of Fisheries (now the Fish and Wildlife Service) in 1933.

could not purchase meat. A discussion on fish buying, inspection, and cleaning conducted by the U.S. Fish and Wildlife Service at a large Boston university in 1948 disclosed that the attending home economics students possessed but limited knowledge of leading New England species.[60] Moreover, many homemakers have never become acquainted with most of the interesting and palatable methods of serving fish. A *Parents' Magazine* survey in 1946 revealed that more than 70 per cent of the women interviewed served only fried fish; 85 per cent asked for instructions on how to thaw frozen fish.[61] The industry might reap large returns if it could meet the challenge to educate those engaged in food preparation, particularly the housewife.

Until after the second World War, New England interests expended little time or money to publicize their wares. The Federal Trade Commission found in 1945 that the aggregate cost of advertising for New England wholesalers and retailers was only a little over one-third cent for each dollar spent by the consumer.[62] Nevertheless, demand grew vigorously between 1935 and 1946. Apparent consumption of fresh and frozen fillets in the United States increased from 115.6 million pounds in 1935 to 257.2 million pounds in 1946, an increase of over 120 per cent.[63] Moreover, fresh and frozen fish [64] originating from the New England and Middle Atlantic Areas [65]

[60] Cf. "Fishery Products in Home Economics Education" (Educational Bulletin 81), *Fishery Products Report 136*, U.S. Fish and Wildlife Service, Boston, Mass., July 14, 1948.

[61] Cf. "Fresh and Frozen Fish Questionnaire," *Parents' Magazine*, New York, March and April, 1946. The findings of Stolting and Garfield agree almost completely with those of the magazine on methods of cooking.

[62] Cf. *Report of the Federal Trade Commission on Distribution Methods and Costs*, Part VIII, June 30, 1945, pp. 25 and 26.

[63] "Apparent consumption" equals production plus imports. Actual consumption varies slightly from the figures cited because of year-to-year differences in cold storage holdings. See "Effect on the Domestic Fishing Industry of Increasing Imports of Fresh and Frozen Fish," *Report of the Secretaries of State and Commerce and of the United States Tariff Commission in Response to House Resolution No. 174, 81st Congress* (Washington, D.C.: U.S. Government Printing Office, 1949), page 24.

[64] Fresh and frozen fish includes here sales in packaged and in whole, or round, form.

[65] Species include haddock, cod, pollock, hake, and flounders originating from New England, and whiting, originating principally from the middle Atlantic area. See R. A. Kahn, "Some Aspects of Future Consumption of Fresh and Frozen Fishery Products," *Fishing Gazette*, Annual Review Number, 1948.

ranked first to sixth in importance in sales in 49 of 65 large cities throughout the nation surveyed by the U.S. Fish and Wildlife Service in 1946,[66] or 75.4 per cent of the urban centers included in the study. Ten years earlier the same species held first to sixth place in only 65.6 per cent of the 61 cities surveyed.

A combination of factors made possible this demand expansion with a minimum of promotional effort by the New England industry. During the decade before the second World War, technological improvements were made in the packaging and freezing of fillets. Improved transportation facilities and expanded distribution outlets brought fillets to consumers previously unable to buy them.[67] The discovery that midwestern consumers would purchase redfish fillets almost without urging, so favorably did the redfish product compare with fresh water varieties, raised New England's packaged fish sales.[68] Production of redfish fillets increased nearly ten-fold, from 4.32 million pounds in 1935 to 41.82 million pounds in 1941.[69] During the war years, consumption of fillets continued to increase because of the shortage of meats, meat rationing, and heavy purchases by the armed forces.

The New England industry entered the postwar period with some recognition that organized promotion might further broaden its markets. It joined with the entire American fishing industry to form the National Fisheries Institute, a trade association of producers and wholesalers.[70] The Institute's primary function is to make the American public "fish conscious." The organization has, however, encountered considerable difficulty in amassing an advertising budget from members' contributions. The Institute's campaign for a million-dollar advertising budget during 1947–1948 fell far short of its goal. National fishery products promotion remains limited. In 1950, for example, food advertisers used more radio time than any other product

[66] Cf. R. A. Kahn and W. H. Stolting, "Sales Patterns for Fresh and Frozen Fish and Shellfish, 1936 and 1946," *Fishery Leaflet 365* (Washington, D.C.: U.S. Fish and Wildlife Service, December, 1949).

[67] See Chapter II.

[68] See Chapter II.

[69] Cf. *Packaged Fish* (Annual Reports, U.S. Fish and Wildlife Service, Washington, D.C.).

[70] Cf. "N.F.I. Flashes," National Fisheries Institute, Washington, D.C., September, 1946. About fifteen years had passed since the collapse of the United States Fisheries Association, the industry's preceding national organization.

group, but very little of that time went toward informing the public of the merits of eating fish.[71]

Since the second World War, the New England industry has taken its own steps toward making New England fish an all-week, all-year-round food. In 1947, the Massachusetts Fisheries Association distributed thousands of fish recipe books to New England housewives through a Boston newspaper. The Gloucester Fisheries Association, in addition to contributing $40,000 for advertising purposes to the National Fisheries Institute in 1949, also published and distributed 100,000 copies of a cook book which features Gloucester fish recipes. The organization paid for those ventures through a voluntary assessment of five cents for every one hundred pounds of fish bought through the Gloucester selling room by the member dealers. During the summer of 1949, the Gloucester and Boston dealers' associations entertained food editors from national women's publications. The magazines published many favorable articles featuring the New England industry and its products. These activities point the way to improved consumer demand.

The Atlantic Fishermen's Union has watched the dealers' limited market development activities with interest. It has taken no part in them. The union insists that the nation's agricultural price support program should be extended to the fisheries.[72] This would solve the industry's demand problem. The union has argued that this would favor conservation, for trip lengths could be limited; it would promote efficiency by placing a premium on fast boats and good fishermen to get the fish to market; it would promote the marketing of fish because fish caught would be fresher; surplus fish could be distributed free through the government's school lunch program with little waste, and this would help youngsters acquire a taste for fish. The union concedes that extension of price supports to the fisheries might require a vessel licensing system because too many people might wish to "go fishing," but it does not see that development of such a system might be an imposing problem.

[71] Cf. "Food Industry Largest Radio Advertisers," *Market Development Bulletin* #263, U.S. Fish and Wildlife Service, Washington, D.C.

[72] Patrick J. McHugh, Secretary-Treasurer, remarked as follows to the Citizens' Food Committee, Washington, D.C., on October 23, 1947: "The Citizens' Food Committee should request the President of the United States to see to it that the present support price system now be extended to the fisheries."

We feel that the program suggested by the union clearly is not in its best interest. It shows, however, how much aware of the importance of demand are the union officials. Unfortunately, the "rosy" picture which seems to follow from the union's attractive presentation of the price support device is really an illusory one. In view of the current market pressures, the program might well put the union as well as the industry at the mercy of the government, for any arbitrary decision to cut the subsidies involved in the policy advocated might find the industry a hapless victim. Moreover, the union's argument overlooks the fact that price supports for fish might wipe out whatever current advantage the fishery industry derives from agricultural price supports which hold up the cost of meat to the consumer. It seems doubtful, then, that the favorable results predicted would follow inauguration of fishery price floors. On the contrary, installation of a fishery price support program would seem fraught with the danger of undermining seriously the New England industry's will to self-improvement. In the face of price guarantees there would be little urge, for example, to improve quality or to bend every effort to extend fish markets.

DISTRIBUTION AND PROFIT MARGINS

Fishery products move through a variety of market channels from the New England fishing centers to the final consumer. Exhibit VIII diagrams the chain of markets from the ocean to the retail level. Dealers and chain stores are the principal purchasers of fish at the ports. They may process the fish into fillets or sell them whole.[73] The larger dealers distribute fish through their own brokers in key cities to jobbers who supply the retail trade. Others sell through independent brokers to wholesalers and retailers. Some dealers ship directly to institutional buyers and retail stores, occasionally on a "cash and carry" basis. They also supply "secondary" wholesalers and processors who move fish to the retail dealer. Some chain stores handle the fish from the pier to their own retail outlets.[74]

---

[73] Sale of whole fish is called "sale in the round."

[74] Notably the A & P Tea Co. and the First National Stores Inc. Before 1947, the latter bought fish from dealers on Boston Fish Pier and acted as a "second-

The available evidence indicates that in the past "charge what the traffic will bear" was the only price policy adopted by the fish dealers. Companies in the major ports aimed at a gross profit of one to two cents per pound on whole fish and a ten to twenty per

EXHIBIT VIII

THE CHAIN OF MARKETS FOR NEW ENGLAND FRESH AND FROZEN FISH

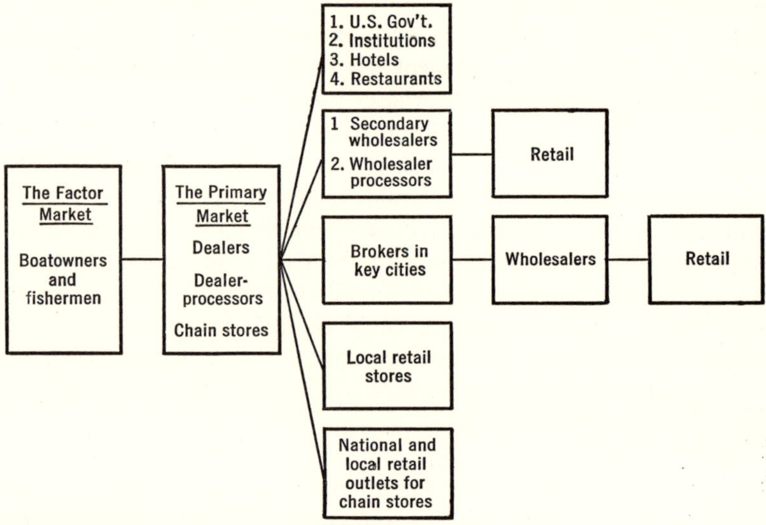

cent markup on fillets.[75] These margins were only "rules of thumb" and frequently are not realized. As one Portland dealer pointed out, he might sell fish that cost him eight cents per pound for as little as eight cents today; next week fish costing him the same amount might bring as much as fifteen cents per pound. In practice, therefore, dealers sold at the highest markup obtainable under existing market conditions in the hope that the average margin over time

---

ary wholesaler." For further discussion of marketing channels, see *Report of the Federal Trade Commission on Distribution Methods and Costs*, part VIII, June 30, 1945.

[75] See testimony of T. A. Fulham, Fulham Bros., Inc., Boston, and F. M. Bundy, President of Gorton-Pew Fisheries Co., Ltd., Gloucester, in *Commonwealth of Massachusetts v. Patrick McHugh et al.*, Suffolk Superior Court, Boston, Mass., June, 1947, pp. 804–856 and pp. 964–967.

would approximate the "ideal level." Few data are available to show dealers' net returns from these margins. A Federal Trade Commission study of operating costs and expenses for two New England primary wholesalers, covering the years 1941 to 1943 inclusive, however, revealed that net profits per dollar of sales averaged only 1.54 per cent of net sales.[76]

Independent wholesalers and retailers seem to have enjoyed the greatest bargaining power in the chain of markets from the ports to the consumer. Wholesalers, who receive detailed information daily on prices paid for fish by the dealers, could take advantage of the high degree of competition among the latter.[77] As the head of the largest Gloucester firm expressed it, "I know of no business that more quickly reflects the changes in supply and demand than the fish business. . . If we have bought our fish at higher prices and the market drops, it is not twenty-four hours before one of our competitors is quoting our customer a price based on the new market price at the boats." [78] In the case of frozen fish merchandising particularly, the wholesalers would seem to have had the upper hand over the main port dealers, for the wholesalers knew from published reports the inventory position of the dealers and the prices the dealers were paying at the dock. Moreover, the wholesalers had their own inventories against which they might balance their purchases; they were closer to, and had more knowledge of, retail conditions. Some idea of wholesalers' gross markups in the past may be obtained from data available for 1943.[79] In that year wholesalers marked up fresh fish about 25 per cent, fillets about 35 per cent, and frozen whole fish about 25 per cent.[80] Satisfactory informa-

[76] Cf. *Report of the Federal Trade Commission on Distribution Methods and Costs*, Part VIII, June 30, 1945, pp. 22 and 23.

[77] The Market News Service, U.S. Fish and Wildlife Service, distributes detailed price and landings information daily from offices at the principal ports.

[78] See testimony of F. M. Bundy, President, Gorton-Pew Fisheries Co., Ltd., Gloucester, in *Commonwealth of Massachusetts v. Patrick McHugh et al.*, Suffolk Superior Court, Boston, Mass., June, 1947, page 804.

[79] Cf. *Fishery Resources of the United States*, Senate Document No. 51, 79th Congress, 1st Session, March 1945, page 127.

[80] Data for fresh fish and for frozen variety cover secondary wholesaling (the third stage in Exhibit VIII). Data for fillets are reported only for "wholesaling" without distinction as to whether primary or secondary. Processing data are reported separately, however, so it is reasonable to presume that "wholesaling" here refers to the third stage in Exhibit VIII. The data in all cases are reported

tion is not available to determine wholesalers' net returns from these markups, but wholesalers handling and financing costs have absorbed a large portion of gross returns in the past.[81]

Retailers have operated on gross margins of 70 to 100 per cent, according to a spokesman for Boston dealers.[82] He hastened to add, however, that many stores have not wanted to handle the product because it is such a cost item. Dangers of perishability loss, one-day-a-week markets, and the inconvenience of handling fish undoubtedly accounted for the superior gross margins of retailers.

Now that the New England industry increasingly is becoming a frozen foods operation, the principal justifications for high markups, particularly at retail, are slipping away. In early 1950, a road salesman for one of the major New England concerns told the author, after the former had completed a swing around the country, that wholesale and retail markups were shrinking and frozen food marketing seemed the major factor at work. At the same time, the New England dealers and processors have come to seek price stability. In the period April through August, 1952, for example, the wholesale price of small haddock fillets per pound quoted at Boston moved in the narrow range of 25 cents to 23 cents per pound. For three successive months it remained the same.[83]

Despite these developments, New England fish continue to cost the consumer about double, on the average, what they bring the main port dealer. In November, 1950, frozen haddock fillets selling at Boston dealers' shops for 23 cents per pound [84] were retailed in

---

in value added terms. We have converted them to their percentage markup equivalents.

[81] Cf. *Defining American Fishery*, Hearings, Committee on Merchant Marine and Fisheries, U.S. House of Representatives, 76th Congress, 3rd Session, April 16, 1940, page 11.

[82] See the testimony of E. H. Cooley, former secretary of the Massachusetts Fisheries Association, in "Meat," part I, *Food Shortages*, Hearings, Special Committee to Investigate Food Shortages, U.S. House of Representatives, 79th Congress, 1st Session, page 369.

[83] See Wholesale Price Data, collected by the U.S. Department of Labor, Bureau of Labor Statistics, and reported in the *Fishery Products Reports*, U.S. Fish and Wildlife Service Market News Office, Commonwealth Pier, Boston, Mass.

[84] Cf. "Wholesale Average Prices and Indexes for Fishery Products," *Fishery Products Report 1*, U.S. Fish and Wildlife Service, Boston, Mass., January 2, 1951. The data were collected and released by the U.S. Department of Labor,

56 large cities at an average price of 50 cents per pound.[85] Redfish fillets, bringing 26 cents per pound in Gloucester,[86] cost the consumer an average of 43.7 cents per pound in the same 56 urban centers.[87]

The power of the New England industry to effect reductions in distributive margins and costs is severely restricted. Only a few of the chain stores, which have fully integrated facilities from the port to the retail level, can maintain control over the later stages of the marketing process. Nevertheless, steps taken to promote a steady, high demand for fish would produce greater competition among wholesalers and retailers to handle New England's product. This in turn should exert downward pressure on wholesaler and retailer margins and pay substantial dividends to New England interests.

Foreign Competition

Competition from abroad is not a new problem to the New England fisheries. New England lost most of its salt fish trade to cheaper producing foreign interests before the first World War. The latter took over eventually even the lion's share of the United States salt fish market, despite duties amounting to 25 per cent ad valorem.[88]

Consequently, New England interests have continuously had a "tariff conscious" attitude. Import regulations have protected fillets

Bureau of Labor Statistics in connection with a new index of wholesale fishery products prices first released by the Bureau on April 12, 1950.

[85] Cf. "Retail Food Prices by Cities, November 15, 1950," U.S. Department of Labor, Bureau of Labor Statistics, Washington, D.C., page 52. The Bureau began only recently to publish these retail prices. They were collected as supplemental data to be used in the revision of the Consumers' Price Index compiled by the Bureau.

[86] "Ocean Perch" is the trade name for redfish fillets, as mentioned before. For source of the above figure, see footnote 84 above.

[87] For source, see footnote 85, above. In 1952, distributors and wholesalers marked up frozen fishery products 17 per cent. This average was second highest among seven frozen food categories. See survey by *Quick Frozen Foods*, reported in *Fishery Products Report 40*, U.S. Fish and Wildlife Service, Boston, Mass., February 27, 1953.

[88] E. A. Ackerman, *New England's Fishing Industry*. Gloucester is active in the salt fish processing industry left in the United States. For over fifteen years the port's supply of salt fish (principally cod) has been imported for processing and packaging from lower cost producing areas outside the United States.

ever since they became a major fishing product.[89] Under the Tariff Act of 1922 the import duty on all fillets was 2.5 cents per pound. The Tariff Act of 1930 continued that rate until the second trade agreement with Canada, effective January 1939, reduced the duty on groundfish fillets to 1.875 cents a pound.[90] The new, lower rate applied to annual imports of 15 million pounds or 15 per cent of the average annual United States consumption of groundfish fillets in the three preceding years, whichever was the greater. Imports of groundfish fillets in excess of the quota and imports of all other fillets remained subject to duty at 2.5 per pound.

An indication of the potential threat to New England interests from foreign fillet producers came in 1939. The largest New England filleting concern made an unsuccessful attempt to establish a new plant in Newfoundland and ship fillets into the United States duty free.[91] It failed only because Congress redefined the "American fishery" to make the products of such a venture subject to the American tariff.[92] This episode not only demonstrated the importance of the tariff barrier for the maintenance of New England's position under the existing cost and marketing conditions, but also it made clear the potentialities for the development of fresh-and-frozen fish facilities in the countries to the north.

Wartime and postwar developments made the challenge from foreign fisheries more acute. During the hostilities, Canada, Iceland, and other northern nations expanded their frozen fillet output greatly to provide fish for the allies.[93] At the war's end, Great Britain an-

[89] All tariff information in this paragraph taken from U.S. Tariff Commission, "Second Trade Agreement between the United States and Canada," volume III, part 2 (Washington, D.C., 1938), page F7–57.

[90] Groundfish include haddock, redfish, flounder, cod, whiting, pollock and hake, as described in Chapter II.

[91] Cf. Hon. R. O. Brewster, "American Fisheries Threatened by Millions of Pounds of 'Duty Free' Newfoundland Fish," Speech in the House of Representatives, *Congressional Record*, February 6, 1940. See also *Defining American Fishery*, Hearings, Committee on Merchant Marine and Fisheries, U.S. House of Representatives, 76th Congress, 3rd Session, April 16, 1940.

[92] Cf. Public Law 600, 76th Congress, 3rd Session, 1940, *U.S. Statutes at Large*, vol. 54, part I, page 305.

[93] New England interests charge that the above nations expanded their filleting and freezing facilities with "lend-lease" funds so that the United States literally "sold the New England industry down the river." Canada, at least, denies having expanded facilities during the war and denies having received "lend-lease" funds for the purpose.

nounced that she would curtail sharply purchases of frozen fillets from Canada, Iceland, and Newfoundland. Britain had bought over 91 million pounds of fillets from the latter nations in 1945, according to a New England spokesman.[94]

Prizing American dollars, these producers turned to the American market. Price inflation facilitated the shift, for it had cut the ad valorem equivalent of the American tariff from a level of 32.9 per cent in 1939 to about 12 per cent in 1946.[95] Foreign producers rushed to take advantage of the situation.[96] The growth in groundfish fillet imports is shown in Table 22. They were over nine times as large in 1951 as in 1939. Imported fillets amounted to 39 per cent of United States fresh and frozen packaged fishery production in 1946; by 1951 imported fillet volume reached nearly 60 per cent of this country's frozen fish output.[97] Thoroughly alarmed by the growing flood of imports, representatives of the New England industry have visited regularly in Washington to seek greater protection. The Fishermen's Union, working through the American Wage Earners' Protective Conference of which it is a member, has in effect spearheaded these drives.[98] The dealers' and vessel owners' associations and the shoreworkers have pooled resources with the Fishermen's Union to try to set an import quota or increase the tariff rate, or both. As we have noted previously, these groups, particularly the dealers and fishermen, have seldom seen eye-to-eye on any other problems. So far their efforts have succeeded only in maintaining present import

[94] See testimony of L. J. Hart, Secretary of the Gloucester Fisheries Association, in *Price Ceilings, Imports and Tariffs on Fisheries*, Hearings, Committee on Merchant Marine and Fisheries, U.S. House of Representatives, February 26, 27, and 28, 1946, page 11.

[95] Cf. *Problems of the Fishing Industry*, Hearings, Subcommittee on Fisheries an Wildlife Conservation, Committee on Merchant Marine and Fisheries, U.S. House of Representatives, 81st Congress, 1st Session, February 15 and 16, 1949, page 20. The percentages represent the relationship of import duty to import prices.

[96] Canadian interests insist, however, that the price inflation hurt that nation's vessel owners by increasing greatly the cost of construction of new vessels to replace those lost in the second World War. (View expressed in letter from Clive Planta, Secretary-Manager, Fisheries Council of Canada, Ottawa, Ontario, Canada, December 14, 1950.)

[97] Data obtained from U.S. Tariff Commission.

[98] This organization, located in Washington, D.C., represents about 500,000 wage earners in about a dozen AFL affiliated unions.

TABLE 22

GROUNDFISH FILLETS, FRESH AND FROZEN: UNITED STATES IMPORTS,
BY PRINCIPAL SOURCES, 1939–1951

Year	All countries	Canada	Newfoundland and Labrador	Iceland	Norway	Other countries
Quantity (thousands of pounds)						
1939	9,426	9,367	46	13	...	...
1940	9,740	9,653	23	47	...	17
1941	9,931	9,744	187	...[a]	...	...
1942	16,674	15,906	728	40	...	...
1943	16,323	14,567	1,084	672	...	...
1944	24,546	21,336	2,529	681	...	...
1945	43,169	37,830	3,937	1,402	...	...
1946	49,260	39,428	5,503	4,323	2	4
1947	35,093	25,909	5,018	4,166	...	...
1948	53,964	34,719	14,875	3,965	396	9
1949	47,322	30,676	11,000	5,131	506	...
1950[b]	64,893	49,212[c]	...[c]	12,767	1,809	1,105[d]
1951[b]	87,639	57,585[c]	...[c]	24,518	4,240	1,296[e]
Foreign value (thousands of dollars)						
1939	714	709	4	1	...	...
1940	853	847	2	3	...	1
1941	964	951	13	...[f]	...	...
1942	2,337	2,257	78	2	...	...
1943	2,706	2,423	174	109	...	...
1944	4,914	4,256	498	160	...	...
1945	8,658	7,398	837	423	...	...
1946	9,951	7,788	1,207	955	...[f]	1
1947	6,193	4,546	884	763	...	...
1948	10,752	6,953	3,120	586	91	2
1949	8,741	5,727	2,101	801	112	...
1950[b]	11,754	[e]9,485	...[c]	1,843	267	159[d]
1951[b]	16,741	[e]11,424	...[c]	4,271	831	215

Source: Compiled by the U.S. Tariff Commission from official statistics of the U.S. Department of Commerce.

[a] Less than 500 pounds.

[b] Preliminary.

[c] Statistics for Newfoundland and Labrador included with Canada.

[d] Includes 625,000 pounds, valued at 105,000 dollars, from Denmark, and 384,000 pounds, valued at 39,000 dollars, from Greenland.

[e] Includes 466,000 pounds, valued at 54,000 dollars, from Germany, 226,000 pounds, valued at 48,000 dollars, from the United Kingdom, and 221,000 pounds, valued at 53,000 dollars, from the Netherlands.

[f] Less than $500.

restrictions from further cuts under reciprocal trade agreements. The evidence which they have presented to governmental officials, however, shows the vulnerability of the New England industry.[99]

New England's principal foreign competitors have several important cost advantages. They are closer to their fishing grounds and can land fresher, higher quality fish. Foreign fishermen can in some instances reach their fishing grounds in as little as three to six hours. The New England fleet, in many instances, sails three days or more to get supplies. On the larger New England vessels, trips last nine

TABLE 23

GROUNDFISH: AVERAGE ANNUAL EX-VESSEL PRICES OF COD AND HADDOCK IN MASSACHUSETTS[a] AND NOVA SCOTIA,[b] 1949–1951

(In cents per pound)

	Cod [c]		Haddock [c]	
Year	Massachusetts	Nova Scotia	Massachusetts	Nova Scotia
1949	6.8	3.9	7.8	5.0
1950	7.2	3.3	8.5	5.6
1951	8.2	3.3	8.8	4.7

Source: U.S. Fish and Wildlife Service and Fishery Section of Dominion Bureau of Statistics (Canada) as reported and compiled by the U.S. Tariff Commission.

[a] Boston, Gloucester, New Bedford, and Cape Cod ports.

[b] Includes North Sydney, Halifax, and Yarmouth.

[c] Prices are weighted averages for all sizes, entrails removed.

to twelve days. Foreign dealers pay 40 to 50 per cent less for groundfish than New England dealers pay. Fishermen in Canada, New England's principal competitor, have been unorganized since their disastrous strike in 1947.[100] Shore labor, too, is unorganized. Hourly wage rates of Canadian fillet plant employees have been on the average 35 to 60 cents per hour or more below prevailing wage rates in

[99] See the Briefs presented by the Massachusetts Fisheries Association, the Gloucester Fisheries Association, the New Bedford Seafood Producers' Association, and the Atlantic Fishermen's Union to the Committee for Reciprocity Information, U.S. Department of State, 1946 and 1948; also the data submitted on their behalf to the Subcommittee on Fisheries and Wildlife Conservation of the Committee on Merchant Marine and Fisheries, U.S. House of Representatives, February, 1949 by America's Wage Earners' Protective Conference (AFL Organizations), Washington, D.C.

[100] This labor dispute accounted mainly for the sharp drop in Canada's 1947 exports to the United States (see Table 22).

the main New England ports.[101] These lower wages in the Canadian Maritime Provinces and Newfoundland reflect the relatively undeveloped economies of those areas. In many cases it is almost "fish or nothing" as their people "simply have no alternative economic activity."[102] Moreover, shoreworker productivity is not sufficiently higher in the New England centers to offset the higher wages paid there.[103]

TABLE 24

COMPARISON OF HOURLY WAGE RATES OF FILLETING PLANT WORKERS IN
CANADA AND NEW ENGLAND, 1951

| | 1951 | | | |
| | Canada | | New England | |
Classification of worker	Newfound-land	Nova Scotia and New Brunswick	Boston	Gloucester
Filleters, cutters, and knifemen	$0.85	$0.95 [a b]	$1.56	$1.45
Packers, male	.81	. . . [a]	1.15	. . .
Packers, female	.46	0.67 [a b]	1.15	0.95
General help	.81	.88 [a b]	1.25	1.25

Source: Rates for Newfoundland from Consular report No. 97, St. Johns and Nova Scotia, Canadian Department of Labor, Ottawa; United States rates from Bureau of Labor Statistics, as reported by the U.S. Tariff Commission, which is the source for this table.

[a] Official data not available.

[b] Derived from data submitted in the *Statement of the Fisheries Council of Canada*.

Despite their greater distance from American markets, foreign concerns have not faced an important cost disadvantage in transporting their fishery products. Water shipment rates are low, and

[101] Cf. "Effect on the Domestic Fishing Industry of Increasing Imports of Fresh and Frozen Fish," *Report of the Secretaries of State and Commerce and of the United States Tariff Commission, in Response to House Resolution No. 174, 81st Congress* (Washington, D.C.: U.S. Government Printing Office, 1949), pp. 30–33. See also Table 24. The fringe benefits of New England workers make the contrast even sharper.

[102] Quotations from "Fisheries in the Economy of North America from the Canadian Viewpoint," address by Clive Planta, Secretary-Manager, Fisheries Council of Canada, to the Boston Conference on Distribution, Boston, Mass., October 20, 1947.

[103] Physical productivity is output per man hour. Opinion obtained from Mr. Joseph Cahill, former Vice President, General Seafoods Division, General Foods Corporation, which operates a filleting plant in Halifax, Nova Scotia. Mr. Cahill expressed this opinion in 1948 before leaving that firm.

rail rates from the maritime provinces to the midwest are about the same as those paid by New England dealers.

Foreign interests, in coöperation with their governments, have plans to expand their sales in the United States market. Canada has undertaken a campaign of fisheries development.[104] Canadian spokesmen deny the charge of New England producers that subsidies to the industry play an important role in the program.[105] Nevertheless, governmentally provided technical services, fish inspection, and industrial aid for low income areas are part of Canada's program aimed at expansion of North American markets. Moreover, that nation's Atlantic coast trawler fleet is growing rapidly as a result of the easing of trawler licensing restrictions.[106] Iceland, Norway, West Germany, and Scotland also have plans for expanding frozen packaged fish sales in the United States. One of these nations — Iceland — particularly bears watching. Exports from that country have expanded faster in the postwar period than those of any other. The Icelandic specialty is ocean perch (redfish) and poses a particular threat to the Gloucester segment of the New England industry. In 1951, that nation alone exported over 10 million pounds of redfish to this country.[107] It has a large fleet of trawlers built in the postwar period; processing plants have been renovated and expanded with the use of Marshall Plan funds; and untapped supplies of fish lie ready for exploiting. That country may be expected to push United States sales hard, too, because over 90 per cent of its exports are fishery products and many of the goods consumed domestically must be bought abroad.

In 1946 and again in 1949 the federal government turned down

[104] Cf. *Trade News*, Department of Fisheries, Ottawa, Canada, November, 1949.

[105] For the New England industry's charges, see *Problems of the Fishing Industry*, Hearings, Subcommittee on Fisheries and Wildlife Conservation, Committee on Merchant Marine and Fisheries, U.S. House of Representatives, 81st Congress, 1st Session, February 15 and 16, 1949. Canadian interests admit that subsidies are granted to encourage new vessel construction, but they insist that the subsidies are not sufficient to offset the advantages of importing fishing vessels from the United Kingdom and the United States. (Information from letter of Clive Planta, Secretary-Manager, Fisheries Council of Canada, Ottawa, Ontario, December 14, 1950.)

[106] Cf. "Canadian Trawler Fleet Expanding," *Fishery Products Report No. 17*, U.S. Fish and Wildlife Service, Boston, Mass., January 24, 1951.

[107] Data from the U.S. Tariff Commission.

the New England industry's plea for greater protection.[108] Finally, in 1951 the industry received a full hearing under the Escape Clause provisions of the Trade Agreements Extension Act of 1951, on the ground that imports threatened serious injury. In connection with those proceedings, the United States Tariff Commission conducted one of the most exhaustive inquiries into the economic features of the industry in the United States that has ever been made.[109] Then, in late 1952, the commission members voted, by the narrow margin of three to two, not to grant the New England industry's request.[110]

The members of the commission voting to deny the requested tariff relief pointed to the generally higher level of economic activity in the industry as of 1951 when compared with the prewar period. They pointed to the vigorous growth of demand which we have noted on pages 118 and 119. They found that "the facts in this case do not indicate that there is any more reason for finding that serious injury to the domestic industry is impending by reason of increased imports than there is for finding that such increased imports are causing serious injury to the industry."[111] The dissenters, on the other hand, stressed their belief that imports will continue to mount at an accelerated rate so that without additional restrictive measures an absolute decline in domestic operations will occur inevitably and soon. They could see little possibility of sufficient market expansion to allow sales of domestic packaged fish at prices which will sustain the industry.[112]

One point emerges very clearly from the flood of facts and claims which piled up during the commission's investigations. Severe and growing competition is an imposing challenge to New England fish-

---

[108] Cf. "Effect on the Domestic Fishing Industry of Increasing Imports of Fresh and Frozen Fish," *Report of the Secretaries of State and Commerce and of the U.S. Tariff Commission in Response to H.R. No. 174, 81st Congress* (Washington, D.C.: U.S. Government Printing Office), 1949, especially pp. 63–68.

[109] Cf. "Summary of Information Obtained During the Investigation on Groundfish Fillets," *Report on Escape Clause Investigation under Section 7 of the Trade Agreements Extension Act of 1951* (Washington, D.C.: U.S. Tariff Commission, September, 1952).

[110] "Groundfish Fillets," *Report on Escape Clause Investigation*, U.S. Tariff Commission, September, 1952.

[111] "Groundfish Fillets," page 22.

[112] "Groundfish Fillets," pages 24–38.

ing interests to settle their internal difficulties, achieve greater efficiency, and develop their markets. Otherwise, the industry will have to concede the correctness of the Canadian hypothesis that New England has reached its present stage of development only through expansion of operations beyond economic limits. The Canadians view the depletion of the New England fishing grounds, for example, as an obvious effect of overfishing due to overexpansion. They insist that New England's problem is to adjust gracefully to lower, long-term production levels based realistically upon locational advantages and disadvantages.[113]

The press of imports should constitute a special spur to the Atlantic Fishermen's Union, for there is increasing evidence that the dealers can and will by-pass the fishermen. The largest dealers in New England have already acquired extensive processing facilities in Canada. Not only are New England interests investing increasingly in foreign operations, but also domestic processors are importing growing quantities of groundfish fillets for sale under their own label in domestic markets. At least one Boston firm has formed a corporation especially for this purpose. The largest independent dealer in Newfoundland already packages nearly four-fifths of his output for New England dealers. They in turn market it in the United States through their well established distribution channels. Further evidence pretty well summarizes the situation. In 1949, Boston dealers handled about 11 million pounds of such fillets, principally from Canada. By 1951 this figure had risen to 20 million pounds.[114] Indeed, it is perhaps not too far-fetched to speculate that before many years pass, the union may find itself quite alone if it talks in terms of increased tariff protection.[115]

One other development should serve to reinforce the spur to the union. Since the fall of 1951, the latter has had an organizer in Hali-

[113] These views were expressed in a letter to the author by Clive Planta, former Secretary-Manager, Fisheries Council of Canada, on December 14, 1950.

[114] Data were furnished through the courtesy of Mr. Thomas Risoli and Mr. Paul Paradis, Boston Market News Service of the U.S. Fish and Wildlife. The figures are estimates from weekly import sheet summaries on file in that office.

[115] We should note that the 1951 plea of the New England industry split the American fishing industry asunder. Wholesalers opposed vociferously any change in tariff conditions. See R. E. Steele, "Brief of the American Importers and Distributors of Imported Groundfish Fillets" (Washington, D.C.: Press of Byron S. Adams), Jan. 5, 1952.

fax, Nova Scotia. While the union claims that the fishermen in that port are somewhat more amenable to organization than those in Portland and Rockland, Maine, union officials nevertheless admit that effective unionization abroad will require a sustained, expensive, long run effort. Up to the end of 1952 a few hundred Canadian fishermen had signed union cards. This picture presents quite a sharp contrast to that in the Massachusetts ports where for many years the union has enjoyed a position of strength and relative security. The union would indeed seem to have the strongest stake of any of the major groups in the principal New England ports in keeping as large a part as possible of current operations intramarginal.

What would be the consequences should the parties' pleas for additional protection be heeded? The chances are, if we judge from the past, that such a development might remove the motivation for industrial improvement. To the extent that this happened, more tariff protection would in effect permit the continuation of uneconomic capacity at the expense of the American public. Furthermore, it would interrupt the flow of resources, particularly capital, to more efficient employment on northern shores.

On the practical side, it would seem that increased import protection would run quite counter to the objective of coöperation between the several nations involved, particularly Iceland and Canada, in administering efficiently under the International Convention the Northwest Atlantic fishery resource. If the latter is to be subject to international control, then it would seem that relatively free trade in fish would be an almost natural concomitant among the nations immediately concerned.

CHAPTER VII

MAKING ORGANIZATION COUNT

THE STORY UNFOLDED in the foregoing chapters is far from a completely happy one for New England fishing interests. Beset by internal conflict and depletion-caused rising costs, the industry faces stiff competition in American markets from meat and the growing flow of fresh-and-frozen packaged fish imports. The future may be bleak unless these difficulties are met. They present a special challenge to the Atlantic Fishermen's Union and the main port dealers to discover a basis for settling their internal difficulties and overcoming their common problems.

Could the union lead the way? Its formation in 1937 marked the opening of a new era in the principal New England fishing centers. The organization commands the allegiance of nearly all fishermen in Boston, Gloucester, and New Bedford, and has given the fishermen a powerful voice in the industry's affairs. It is the one organization which effectively embraces all three ports.

Moreover, the profit-sharing wage lay provides the union and its members with a unique stake in the industry's managerial efficiency. Favorable and unfavorable turns in the industry's well-being affect the fishermen immediately and directly. As we have seen, the lay ties their fortunes to the production and marketing performance of the industry.[1] High costs mean high costs to the fishermen, for they pay the operating charges of each fishing trip; poor sales mean lower values to the fishermen for their product. Conversely, the fishermen unlike most "ordinary" wage earners, can share immediately the fruits of improved sales and lower operating costs.

The basis for constructive union leadership is, therefore, the important and sound one of enlightened self-interest. Successful development of a forward-looking program would give more job protec-

---

[1] See Chapter V.

tion and income benefits to individual union members. It would build union strength, not weaken it.

Past experience indicates clearly that the approach needed is not one of pious aspirations to coöperate. The 1946 Boston agreement represented such an attempt. The dealer-owners and the union pledged solemnly to forget past conflicts and to coöperate wholeheartedly. No such transformation occurred.[2] Neither did it help much that the new executive secretary of the Boston employer associations enjoyed much more congenial relations with the union leadership than had his predecessor.[3] Building for progress requires a solid foundation erected through hard, realistic bargaining by the highly skeptical parties.

The improvement of primary market price-making arrangements is the first most essential move. The worst and most persistent clashes between the fishermen and dealers have concerned the processes of determining catch values.

How might the union initiate favorable changes in dealer-fishermen relationships? What demands should the union make? Equally important, what concessions might it offer? The specific moves which might be desirable differ in Boston, on the one hand, and Gloucester and New Bedford on the other, because of varying conditions in the respective centers. For this reason we shall discuss separately the internal conflict problems of each port.

Resolving the difficulties of determining the value of the catch is the most basic but not the only task. There remain the perplexing problems of depletion, marketing, and foreign competition. They concern all the principal groups in all the ports, including the relatively unorganized centers of Portland and Rockland, Maine.[4] How might the union spearhead a joint effort to deal realistically with these problems? In what ways might the federal and state governments aid the industry to help itself?

The Foundation for Coöperation

The union's best opportunity lies in Boston, where the market organization makes all difficulties matters for collective bargaining. As we have noted, the fishermen and dominant dealers in that port

---

[2] See Chapter V.   [3] See Chapter IV.   [4] See Chapter IV.

are related both as sellers and buyers and as employees and employers, since the leading dealers own most of the vessels operating.[5]

Boston's difficulties have arisen mainly from the operation of the sellover feature of the New England Fish Exchange Rules.[6] The dealers insist that the sellover accomplishes its objective of making quality count in the determination of final catch values. The fishermen charge that the sellover is nothing more than a "price cutting device built into the market" to depress the fishermen's earnings under the profit-sharing lay. The analysis developed in Chapter V suggests that, as currently constituted, it is both.

Three factors account principally for the sellover turmoil. First, the New England Fish Exchange Rules providing for the sellover are vague and inadequate. They set no determining criteria on which the sellover operation might be based. Hence, under present arrangements, the sellover maneuver is much more likely to come into play because of market forces than objective factors of quality.[7] Second, the arbitration provision of the Exchange Rules for settling quality disputes is most unsatisfactory.[8] The decision is usually up to the Exchange manager, who is mistrusted by the fishermen because he is indirectly an employee of the dealers who make up the Exchange membership.[9] Moreover, the incumbent who has held the office for more than a score of years operates a supply house which sells materials to the dealers. Third, the integration of trawler-owning and fish-dealing in the Boston market tends to make the vessel owners more interested in obtaining cheap fish for processing than in maximizing their profits as vessel owners.[10] The fishermen, of course, want the best possible prices for the fish they land.

The union, on the fishermen's behalf, insists that the answer to sellover problems is to eliminate the resale feature of the market. Union spokesmen argue that as long as the fishermen are convinced that the dealers "steal" fish at lower prices through the sellover device, the latter provides little incentive to turn out a higher quality product.

[5] See Chapter V.
[6] For full description of this process, see Chapter II and Chapter V.
[7] See Chapter V.
[8] See Chapter V.
[9] See, however, footnote 46, page 76.
[10] See Chapter V.

It would be better, however, for the union to concentrate upon obtaining improvement in the sellover operation — upon making it really a quality-promoting feature of the Boston market. Union members as well as the dealers stand to gain from higher quality production.

Through collective bargaining the union might propose that the final arbiter on fish quality be an outside agent acceptable to both the dealers and the fishermen. He should be paid jointly by the parties and should be allowed to hire competent assistants if necessary. This move would help eliminate needless controversy. Secondly, the union might propose that each captain break down on a quality basis the quantities of each species of fish offered for sale in the opening market.[11] It is ridiculous for all fish to be represented as first quality in the opening market, as at present. Almost always there will be some second grade fish in each catch, if for no other reason than that the earliest caught fish on each voyage may be nine to ten days old when landed. These earlier caught fish, moreover, will have been carried hundreds of miles in the vessel's hold under the weight of later hauls with consequent damage to quality.

If this change were made, specific quantities of two grades of fish might be offered in the first sales market. Each captain, of course, would minimize the proportions of his catch which he would tab "seconds," or "No. 2." Fish which he did so classify would, undoubtedly, be "No. 2," barely suitable for sale in fresh or frozen form. The impartial inspector would, then, have two sets of cases to decide. First, he would entertain complaints from the buyers of No. 1 fish that the latter were not as represented. In such cases there would be a sellover whenever the inspector uncovered No. 2 fish. Second, the inspector would follow a similar course whenever a No. 2 fish buyer insisted that the fish offered him were misrepresented. If the buyer's claim were upheld, the fish would be "sold over" as No. 3. As a practical matter this second class of decision would seldom be involved if the present No. 3 concept were retained.

---

[11] The present practice of listing supplies by banks where caught is something of a quality indicator. Georges Bank haddock, for example, are considered better than those from northern grounds.

Were the dealers to accept the union's proposals, the changes would give the fishermen every incentive to land the highest quality of fish possible in order to have them go as No. 1 in the first sales market. This in itself should minimize the proportions of second-grade fish. Provided the captains were reasonably accurate in listing the grades of each species in their respective catches, the "sellover" should shrink to insignificance. The captains would have every reason to be careful in estimating, for they could be sure that quality would be judged fairly. Admittedly, this system would not be as good as one in which all fish were unloaded, graded impartially, and sold as No. 1, No 2, and No. 3 fish, respectively, during the one auction period at the opening of the day's business. But such an ideal would be virtually impossible of attainment. For one thing, the very handling involved in removing and grading would increase the amount of No. 2 fish; for another, there probably would not be the space available to set up the graded fish for sale. The system we advocate would, though, formally recognize two grades of fish in the opening market.

The dealers should have no overwhelming objections to the suggested market changes, for they provide merely that impartially determined quality would play a fuller role in fish sales. The union could, however, induce dealer coöperation on this point by offering to maintain a "spare list" of fishermen who would be available as replacements should some crew members fail to appear at an appointed time before scheduled ship sailings. Such a system could be run conveniently through the union office; it would eliminate "straggler" interference with regular departures for the fishing grounds, a bitter complaint of the dealer-vessel owners. Such a union offer would not run afoul of the Labor Management Relations Act of 1947 because the union would merely be enlisting its coöperation to help achieve full vessel operations. The move would not be one to force vessel owners to employ only union men, even though its practical effect would be to facilitate the union's central responsibility for an adequate labor supply. It is doubtful that the union would have to admit more men to membership to carry out such a commitment. The principal result of the arrangement would most probably be more efficient utilization of the existing manpower supply. Once such a system were operating to supplement present hiring arrangements, the union would be in a position to adjust its admissions to membership realistically in terms of the port's needs.

The foregoing steps could be taken with a minimum of delay. At the outset, the jointly chosen arbiter's judgment could settle quality questions. To perfect the system, development of objective quality standards could be set as a longer range aim. The research laboratories of the U.S. Fish and Wildlife Service are ideally suited to develop such criteria. They are working constantly on special aspects of fish quality problems.[12] If necessary, the union might offer to match a small contribution by the dealers toward the expense of such a project. In Maine, for example, the sardine packers have had such an arrangement for the study of sea herring since 1947.[13] Should the parties prefer a private agency to develop the criteria, one large research organization has already expressed interest in such a project.[14] An important by-product of such studies might be the discovery of quality-protecting techniques of immense value to Boston interests.

There remains the question of market organization and effective competition in Boston. We have noted that the integration of trawler-owning and fish-dealing has tended to tip the market in favor of the buyers.[15] We have seen, too, that the union's reaction has been to stipulate prices at which catches would be sold or to set fleet-wide limitations on production.[16] Neither situation is beneficial to the port; both promote conflict and discord; the restrictionist moves tend to raise costs directly.

The union could encourage healthy competition by contracting with Boston's dealer-owners not to stipulate prices or to enforce any fleet-wide production controls unless the owners were first consulted

---

[12] See for example, S. R. Pottinger, R. G. Kerr, and W. B. Lanham, Jr., "Effect of Re-freezing on Quality of Sea Trout Fillets," *Commercial Fisheries Review* (Washington, D.C.; U.S. Fish and Wildlife Service), January, 1949, pp. 14–16.

[13] The New England sardine business, located principally in Maine, is a separate branch of the fishing industry not connected directly with the fresh and frozen sector here under investigation. The Maine packers contribute four to six thousand dollars a year to help finance research by the U.S. Fish and Wildlife Service and the Maine Department of Sea and Shore Fisheries.

[14] The United States Testing Company manifested such an interest in the spring of 1950. An official of the firm discussed the possibilities with the author, but on investigation decided that attack on the quality problem would best await the initiative of the parties in the industry itself.

[15] See Chapter V.

[16] See Chapter V.

and the reasons for the fishermen's dissatisfaction thoroughly aired.[17] In return, the union might exact from the dealer-owners the guarantee that (1) each vessel would make its own production policy while at sea on the basis of existing fishing and marketing conditions [18] and (2) each captain should decide freely whether to sell his fish himself or to allow a hired commission man to do so.[19] This would bind the owners to let each captain and crew use its own judgment on what species to catch, what quantity to bring ashore, and how long to stay at sea on each trip. Also, each captain and crew would decide how much they should seek for each catch landed. Such an agreement would be easy for the union to supervise, for every crew member would in effect be a policeman.

Both groups would gain from the exchange. The union would be assured of competitive operation of the boats; the dealers would be unable arbitrarily to order their vessels to stay at sea or head immediately for port.[20] With the sellover machinery improvements previously outlined, the union could be reasonably certain that buyers and sellers met on equal terms. In short, the dealers could not engineer a "glutted marked." [21] On the other hand, the dealer-owners would be rid of the cost-raising union restrictions on output which they have long deplored. The dealer-vessel owners could make a real contribution to more stable relations in this context by making sure that the union leaders and members were informed of product market prices, selling obstacles, and what was being done about them by the dealers.

Accomplishment of the suggested changes in the Boston market would silhouette cost and marketing difficulties as the real villains at work whenever the fishermen's incomes tended to stumble or dealer-owners' net returns faltered.

Securing the same results in Gloucester and New Bedford would

---

[17] That is, the consultation with the owners would be to find ways *other than* production controls to solve the problem.

[18] Each vessel has radio and/or telephone; the vessels are in touch with market conditions ashore.

[19] This has been a continuous collective bargaining issue. See Chapter V.

[20] The union has suspected strongly such dealer moves in the past, particularly when several vessels "happened" to appear on the same day in port.

[21] The union has charged such dealer moves in the past, but has never been able to produce sufficient evidence to validate the complaints.

be more difficult. The fishermen and dealers are members of separate economic groups; their representatives have not bargained collectively. Moreover, fish landed in the latter ports are not unloaded at a central pier as in Boston; they are removed at the respective dealers' docks along a half mile or more of waterfront.

Disputes in Gloucester and New Bedford have centered about weight allowances and weighing methods.[22] In the typical instance a given dealer would purchase a "trip" of fish through competitive bidding in the union-run selling room. During the process of unloading, a dispute would arise at the dealer's dock either (a) because the fishermen felt the dealer's scales gave "short" weight, or (b) because the dealer and the fishermen could not agree on the percentage to be deducted from the gross catch weight as an allowance for ice, trash, and poor fish not suited to processing. In some instances both matters would be at issue. The second has been particularly troublesome in Gloucester, a redfish port, for the fleet has had to resort increasingly to capturing small fish in the face of growing scarcities of redfish supplies.[23]

The fish auctions themselves do not seem to be the seat of the problem. Both selling rooms provide full opportunity for all dealers to bid, without charge, on a value-offered basis for the respective vessel hauls up for sale. With the exception noted in Chapter V [24] the union has run but not interfered with the free operations of these markets. The union representatives who conduct the bidding act only as intermediaries between the dealers and the respective captains. Again with the exception cited, the dealers have not complained about selling room procedures. One important reason for this is that the union has not generally opposed improvisations in bidding. The dealers, for example, have been able to offer different prices for a given species on a particular vessel, according to the size of the fish in question or the fishing bank upon which the fish may have been caught. Thus a given dealer may offer three prices for a particular species on a given vessel, one price for small, one for medium, and one for large fish. The dealer's value bid, which is the captain's main

---

[22] See Chapter V.

[23] See Chapter VI. Small fish are difficult and sometimes impossible to fillet because their yield is so poor.

[24] See "Union Policy and Public Policy."

interest, will be determined by multiplying the prices offered by the respective quantities of each size and species listed by the captain as part of his catch. After the sale at auction, however, there may be considerable controversy between the purchasing dealer and the captain or crew because the dealer disagrees about the proportions of small, medium, and large fish in the catch, or for the other reasons we have indicated.

The solution to Gloucester and New Bedford conflicts over the determination of the catch value is not, therefore, to change selling room arrangements. While it might be desirable for the dealers to have representation in the management of the selling rooms (and part of the financial responsibility for their support), such a move is no more necessary than that of giving the Atlantic Fishermen's Union a seat on the Board of Directors of the New England Fish Exchange in Boston.[25] Moreover, adoption of some sort of sellover arrangement would not seem the answer, at least until the Boston situation is cleared up and has a chance to prove its worth. Neither the dealers nor the fishermen have expressed any wish for such an addition. Both groups know of Boston's sellover troubles; they have no wish to become similarly embroiled.

Gloucester and New Bedford need definite machinery to settle matters which may be essentially moot after auctions take place. The ports require an objective, orderly process to insure (a) that the dealer pay the proper price for the correct proportions of differently classified fish in each catch, e.g. small, medium and large, and (b) that the dealer pay only for the correct weight.

The union could spark such a development by insisting that the dealers (a) use bonded weighers wherever allegations of "short weighing" crop up and (b) that the dealers agree to a grievance committee arrangement to settle finally any honest differences of opinion concerning poor fish, ice, and trash deductions. The grievance committee might include (1) the two union delegates in each port, (2) the complaining dealer and any other whom he might select, and (3) an impartial chairman retained jointly by the parties. Decisions of such a group might develop gradually a set of rules so that most disputes eventually might be settled by the contesting parties

[25] In the 1940 negotiations the union sought this concession from Boston's dealer-owners. See Chapter V.

without the committee's intervention. Since there is nothing "monopolistic" in such an arrangement, it would seem that a marketing agreement might be drawn between the union and each dealer to set up the machinery. It would seem, too, that the union legitimately might deny bidding privileges in the selling room to any dealer who refused to observe the agreement's terms.

How might the union get the dealers initially to fall in line with such a proposal? The union has sought at least once in the past to install the use of bonded weighers, but without success.[26]

The answer is that the union must offer the dealers a worthwhile inducement which will at the same time demonstrate clearly that the fishermen wish wholeheartedly to improve the port's marketing conditions. A union offer to give up its policy of instituting unilateral output controls would be just such a concession, long sought by the dealers.[27] The union might agree to leave production decisions in the hands of the individual vessel. The latter is best qualified, anyway, in view of the uncertainties of each fishing voyage, to make such decisions. Until the union makes such a move, it seems clear that the dealers will balk every attempt to set up processes which interfere with unilateral dealer determination of weight, percentage, and so on in finally settling catch values.

The changes suggested for Boston, Gloucester, and New Bedford can be achieved by the respective groups through hard bargaining. Indeed, this is the only way that initial, constructive moves may be made by bitter, skeptical traders. But the resulting arrangements might very well provide a foundation for the growth of more coöperative relationships between union fishermen and the dealers. The "ready made" issues which have kept these diverse interests at war either will have been carved out or will have become amenable to solution through carefully drawn processes. With the union leading, the parties will have erected an "industrial jurisprudence"[28] system to meet their own peculiar problems.

The necessary supplement to these steps is a vigorous attack upon the industry's cost and marketing problems. Better economic health

---

[26] See Chapter V.

[27] The offer might reasonably include the qualification suggested for Boston on page 141.

[28] Cf. S. H. Slichter, *Union Policies and Industrial Management* (Washington, D.C.: Brookings Institution), 1941.

would contribute to the firm growth of unity. In moving to cope with the deleterious influences of fish scarcity, marketing, and foreign competition difficulties, vessel owners and shore workers, as well as the fishermen and dealers would have the common interest of insuring their future. What directions should constructive action take? How might the Fishermen's Union point the way?

A Constructive Program

Coördinated, systematic research, market development, and improved industrial practices are the steps which should provide an answer to the industry's fish scarcity, marketing, and foreign competition problem. Let us examine each of these resorts before we discuss means of implementing them.

RESEARCH

The industry needs an institute to develop, collect, and disseminate knowledge applicable to both technical and economic problems. Technical inquiries might uncover and classify, for example, the merits of species which are now abandoned or underexploited. In turn, marketing research might discover both edible and industrial uses for which those species might be sold. Broadening of the industry's product base would be the highly profitable result. The key species might be relieved of some of the pressure of concentrated fishing efforts. The efficiency of fishing operations would be improved. In recent years as much as twice the amount of fish landed has been discarded at sea as unmarketable for one reason or another.[29] Were more of the species marketable, the New England fleet would enjoy substantially higher output per day of fishing effort; each vessel would utilize its capacity more fully with consequent reductions in unit costs. Shore employment would be bolstered and wider spread of overhead on shore facilities made possible through greater output volume.

Technical research also could help the industry to improve the

[29] Cf. Earl Banner, "Trawler Men Have to Throw Away More Fish than They Bring Home," the *Boston Daily Globe*, July 28, 1949. This article reports the experience of F. W. Sargent, Massachusetts' Director of Marine Fisheries, on a Boston trawler ten-day trip.

quality of its product. Recent studies indicate, for example, that fish stored in the hold of a vessel will deteriorate about twice as fast at 37° as at a 32° temperature.[30] Moreover, learning to reduce refrigeration shortcomings is vital for the strengthening of the chain of quality retention from the sea to the consumer.[31] Research in this area alone can provide handsome dividends. Nothing hurts sales more than spotty quality.

No less important, a research arm might accumulate, organize, and circulate among interests in the industry full information on important technical developments in process or completed by far-flung fisheries agencies throughout the world. The United States Fish and Wildlife Service, through its technical representatives and various publications[32] currently reports such information, but no agency especially interested in New England problems is available to see that the most is made of the information. Any number of advances might be worth adoption by New Englanders, once they were explored and understood. Recently, for instance, German interests have started deep-sea dragging operations with an electrical fishing device. Reputedly it will enable a vessel to sweep from the sea the 90 per cent of the total number of fish which pass between the ship and the net, but which escape orthodox trawls.[33] If adaptable to New England operations, this scheme might increase physical productivity tremendously. The result could be shorter trips, fresher fish, and less wear and tear on crews, boats, and equipment.

Two other salutary developments might follow the founding of a formal research agency. First, the latter could serve to provide a clear channel for coöperation on technical conservation matters between the industry itself and the scientific branch of the Interna-

---

[30] Cf. H. E. Crowther, "Control of Fish Spoilage by Icing and Freezing," *Commercial Fisheries Review* (Washington, D.C.: U.S. Fish and Wildlife Service), March, 1951, pp 6–10.

[31] Cf. G. A. Reay, "Factors in Distribution Affecting the Quality of Fish," *Chemistry and Industry*, December 9, 1950, pp. 789–790.

[32] The Service publishes brief bulletins through its daily Market News Services; articles in its *Commercial Fisheries Review*, a monthly periodical; abstracts in its monthly *Commercial Fisheries Abstracts* release; and special monographs on selected topics in its "fishery leaflets," occasional papers, etc.

[33] Cf. "German Vessel Equipped with Deep-Sea Electrical Fishing Device," *Fisheries Products Report No. 251* (Boston: U.S. Fish and Wildlife Service), December 28, 1950. The device sets up an electrical field between the ship and the net. This stuns the fish so that they are swept into the net.

tional Convention of the Northwest Atlantic Fisheries.[34] The move would be convincing evidence that the industry is anxious to help in enlarging the technical and biological knowledge of the fisheries. Second, the research agency's economic arm, in addition to the marketing research already suggested, could undertake investment, cost, price, and transportation studies. In short, it could aid the industry to ferret out weaknesses and reinforce its strength. Even a modest business advisory service would be invaluable.[35]

MARKET DEVELOPMENT

As important as are the other elements in a forward-looking program, none is more vital than expansion of markets. Improving sales for both edible and industrial uses merits the industry's careful attention. Here lies the answer to the problem of more income and, in part, to foreign competition.

Fish for food deserves primary emphasis, for to date it has provided by far the higher value product. The available evidence indicates that consumer ignorance of the variety of fishery products, the nutritional qualities of fish, and the varied methods of preparing fish blankets a huge slumbering demand. It could be awakened by more widespread education about fishery products, by persuasive advertising, and by attractive packaging.[36] Intensified programs to inform the housewife about the use of fishery products might (a) popularize the industry's lesser known species whose merits were discovered and developed fully by the research activity previously described and (b) make New England fish more and more an all-week, all-year-round food. These achievements would increase competition among wholesalers and retailers to handle fishery products and in turn might scale down distribution margins which seem in the past to have absorbed a large percentage of retail prices. The result would be an increased, steadier flow of income to the New England centers for all interests to share.

The industry should not neglect the promotion of fishery products sales for nonedible and industrial uses. In recent years this has been

[34] See Chapter VI.
[35] See Chapter III for discussion of business practices.
[36] See Chapter V.

one of the fastest growing sectors of the New England fish business. In 1945, the United States production of groundfish meal and scrap, principally a New England product, amounted to 22,471 tons worth $1.57 million.[37] By 1949, production had jumped to 31,425 tons worth $5.22 million. In that year, the demand for fish meal, fish oil, and fish solubles had so increased that large sectors of the Southern New England fishing fleet turned to "trash fishing" — the search for species not saleable in edible form. They landed about 75 million pounds of such fish to supplement supplies of waste and scrap from fillet houses.

The growth of industrial outlets for New England fish presents the industry at once with a challenge and an opportunity. The challenge is to coördinate research and market development so that: (a) the expansion of "trash fishing" will concentrate upon relatively unexploited fish stocks; insuring this will avoid the destruction of young, edible species before they are large enough to market profitably and, at the same time, may benefit the stocks of edible fish on the banks by removing competitors for the food supply; and (b) so that the industry will receive the best value for its edible and nonedible products. It would be folly to exploit intensely for by-products purposes at a cent per pound a species which might, through careful promotion, bring several cents per pound if it were sold for edible uses. This is a serious problem, for many of the so-called "trash-fish" exploited thus far are edible but not popular. Even for those species which must be sold as nonedible, however, it would be to the industry's advantage to discover and develop sales which will provide the product of the highest value. In this connection, the parties might well join together to promote sales of those fish in limited demand — that is, those species which bring a good price, but currently sell only to the few who know their merits. (Indications are that the underexploited pollock is in this category.) The union might agree to land a certain quantity of these species for nothing. Dealers and shoreworkers might also contribute their services. The fish after processing could be shipped at cost of delivery to retailers in selected areas, with the suggestion that fish be sold cheaply with an appropriate "tie in" — soup, sauces, and so on. The industry simply cannot afford to leave future major developments to chance, as it did the redfish expansion

[37] All data from the U.S. Fish and Wildlife Service.

of the 1930's.[38] If the industry will meet the challenge indicated, then its opportunity to expand revenues is wide open.

There is every reason to include foreign interests in these plans. An excellent example of the approach needed was set in the summer of 1951 when Mexican and American shrimp dealers combined to form an international shrimp association. The primary objective of the organization is to take steps to expand the American market. At the time the association started to take form, the American members were still hard pressed by Mexican exports to this country, too.

IMPROVED INDUSTRIAL PRACTICES

Perishability is the central feature of New England fishery operations, for the fish begin to deteriorate the minute that they are hauled from the ocean. It must be remembered that later freezing cannot restore deterioration losses; at best, freezing can but maintain the quality present at the time of freezing.

The industry's key task, in seeking to maintain consistent high quality output, is to preserve the best features of fish fresh from the sea. Indeed, the industry must do this to build sales; adequate promotion and attractive packaging are supplements, not substitutes for clean, delicious, self-selling products.

The process of improving industrial practices must begin at sea. Past investigation indicates that a substantial number of fish spoiling bacteria, including those found in the slime and feces of fish, are of marine origin.[39] They can grow rapidly at temperatures not much above freezing. For this reason, the fishermen can do much to improve quality by careful "gutting," [40] painstaking handling to avoid bacteria-inviting breaks in the fish's skin, and thorough icing and storage. The latter is particularly vital. The fish must be stored so that they get maximum preservation from icing and so that they are

---

[38] See Chapter III.

[39] Cf. C. H. Castell, "The Control of Fillet Contamination in Fish Plants, Part II, the Relationship between the Initial Contamination and the Subsequent Rate of Spoilage," Fisheries Research Board of Canada, *Progress Reports of the Atlantic Coast Stations*, No. 41 (Jan., 1948), pp. 10–14.

[40] "Gutting" is the process whereby the fishermen remove the entrails from newly caught groundfish before icing them down for the trip to port. The trick is to make the removal complete and thorough.

not crushed in transit.[41] Since both owners and union officials agree that the "hold man" — the fisherman who ices down the catch — plays a vital role in preserving quality, they might consider seriously negotiating a bonus for that job on each trip to insure that it is held by a good man. Moreover, eliminating the use of "pitchforks" would be a good first step in better handling of fish aboard the vessel and in unloading.[42]

There is considerable room for improvement, too, in handling fish ashore.[43] As we mentioned earlier, many dealers whose plants are some distance from piers have often left fish in carts exposed to the weather for up to half a day to await processing. When such lay-overs appear necessary, the dealers should make provision for protecting their supplies. At the very least, such fish should be iced down and covered.

Complete sanitation is vital in processing plants. Careful investigation by Canadian scientists has revealed that cod, cautiously filleted to prevent bacterial contamination, kept more than twenty days at 38° Fahrenheit without loss of palatibility or noticeable spoilage. Similarly-cut fillets, exposed to the filleting tables of a "well-run" fish plant for only twenty seconds and then stored in sterile containers at the same temperature, went bad completely in six days.[44] This experience seems to indicate that even the more carefully run, larger plants in New England might benefit from an inspection system which would stress improved shop practices.

Most New England dealers, with the exception of the larger chain stores, lose control of their product once it is shipped.[45] Control of quality passes out of their hands. Nevertheless, much sound counsel on the safest shipping methods to insure high quality maintenance has become available to the industry.[46] The larger New England

[41] Cf. "Stowing Ice Aboard to Improve Fish Cargoes," *Western Fisheries,* vol. 40, no. 2, May, 1950, pp. 54 and 58.

[42] See Chapter II.

[43] See the remarks of Dr. W. F. Hampton, General Seafoods, in the *Atalantic Fisherman,* January, 1952, page 5.

[44] Cf. E. H. Castel, "The Control of Fillet Contamination in Fish Plants, Part II, the Relationship between the Initial Contamination and the Subsequent Rate of Spoilage," Fisheries Research Board of Canada, *Progress Reports of the Atlantic Coast Stations,* No. 41 (January, 1948), pp. 10–14.

[45] See Chapter VI.

[46] See for example W. A. MacCullum, "Changes in Cooling and Transporta-

firms could instruct their road salesmen and brokers to acquaint both wholesalers and retailers with the best ways of merchandising the New England industry's product. It would cost little to do this and would bring those dealers immediate returns through improved sales relationships.

The three basic steps to sound prosperity which we have outlined have been urged upon the industry in one way or another for a number of years. As early as 1919, Lewis Radcliffe of the U.S. Bureau of Fisheries pointed out the merits of research for unsolved commercial fisheries problems.[47] Fishery trade publications and the U.S. Bureau of Fisheries have exhorted the industry to pay greater attention to marketing difficulties.[48] A generation ago, a leading Boston fishery executive began pleading for improved industrial practices.[49] The pitchforking issue has stood out as a discussion topic for at least twenty-five years.[50] The industry's principal groups have found it difficult to achieve that workable coöperation so necessary for coming to grips with their real problems.

The Atlantic Fishermen's Union might lead in changing this state of affairs. The last fifteen years have witnessed the growth of organization among all the principal groups in Boston, Gloucester, and

tion Technique Suggested for the Marketing of 'Fresh' Fresh Fish," *Fisheries Research Board of Canada, Progress Reports of the Atlantic Coast Stations, No. 45*, April, 1949, pp. 11–12; A. L. Reneau, "Frozen Food Transportation and Distribution," *Ice and Refrigeration*, October, 1949, pp. 17–19; and "Shipping of Fish and Sea-Foods," *Fishing Gazette*, October, 1949, pp. 64 and 66.

[47] Cf. Lewis Radcliffe, "Fishery Laboratories Afford the Greatest Promise of Relief of Unsolved Problems Affecting Commercial Fisheries," *Transactions of the American Fisheries Society* Vol. XLIX, No. 1, December, 1919, pp. 3–18. The paper was awarded a prize of $100.00 for the best contribution on the solutions of problems affecting commercial fisheries work.

[48] See for example the following: "The Whiting: A Good Fish Not Adequately Utilized" (Washington, D.C.: U.S. Bureau of Fisheries), Economic Circular No. 32, October 10, 1917. This species acquired some popularity only in the late 1930's as the result largely of straightforward action by O. L. Carr, a leading midwestern fish dealer. See also the following articles from the *Fishing Gazette*: "Eliminate Gluts by 'Demand Markets,' Establish 6-day-a-week Retailing," November 1, 1924; "Advertising Needed to Tell the Public About Fish," *Annual Review Number*, 1935; "Here Is What Other Food Trades Are Doing to Promote Business," February, 1939.

[49] Cf. J. C. Wheeler, "Inspection at the Source," *Fishing Gazette*, November, 1921.

[50] Cf. "Held Up by a Pitchfork," *Fishing Gazette*, August, 1927. See also Chapter II.

New Bedford.[51] Most striking has been the development of solidarity among the fishermen. Their union occupies a strategically powerful position in the fish marketing structure. Its locals in the respective ports are linked closely under Boston's leadership. Having secured the market adjustments proposed earlier in this chapter, the union might turn into constructive channels resources and energies formerly devoted to fighting. Adequate provision for value determination should enable the union to see clearly that unfavorable pressures on the fishermen's incomes would be emanating from cost and marketing disadvantages, not from "dealer squeeze plays." The union could make a singular contribution by using its strength to swing all the industry's organized groups into a united forward march.

In taking the initiative, the union might announce to the industry its willingness to lead in founding a New England Fisheries Research, Development, and Improvement Association. The union could offer a fixed annual sum — say $10,000 — toward financing the move. It might invite the dealer organizations among themselves to match that figure. The several vessel owner associations together might be asked to contribute two-thirds the amount offered by the fishermen's organization — the relationship which the owners' gross share in the lay bears to that of the fishermen. The seafood workers' locals together might contribute the same percentage of the dealers' offering that direct shore labor costs on the average in each port bear to total fillet manufacturing costs. On the basis of 1944 Gloucester data, for example, that local's contribution ratio would be about one-fifth.[52] Such a financing arrangement would be reasonable, for it would tend to reflect the relative importance of the industry's several groups. Representation on the new organization's board of directors might be divided similarly. If the parties were to desire a different system of financing and representation, however, it could be worked out at a series of conferences arranged by the Fishermen's Union.

The organization resulting could have three branches, one for each of the major responsibilities spelled out in the association's title. The association director might coördinate branch efforts and provide

[51] See Chapter IV.
[52] Source: Unpublished Special Cost Study, U.S. Fish and Wildlife Service. Total manufacturing costs amounted to 22.2 cents per pound and labor costs to 4.66 cents per pound.

liaison with the industry governing board. At the outset, the association's modest means would limit it to (1) drawing up a careful inventory of industry strength, weakness, and problems; (2) assembling all readily available information on the above matters; (3) developing and implementing recommendations where possible; (4) gathering full information on research, technical and developmental aids of which the industry might take advantage; (5) acquainting interested agencies, especially the federal and state governments, with particularized situations upon which the industry would welcome aid and coöperation; (5) working out and coördinating arrangements with such agencies for performance of specific helpful projects. As a longer run aim, such an association might develop gradually the capacity to do its own limited research and development tasks.

In line with these objectives, the association could take immediate, practical steps. It could adopt an attractive quality "seal of approval" for the use of dealers and processors whose operations met consistently a set of standards which a subcommittee of the dealer and seafood worker segment of the association might develop with reasonable speed. The seal might be featured (a) in whatever limited promotion the association might be equipped to undertake, (b) in the advertising of the National Fisheries Institute Inc.,[53] on behalf of its affiliated New England membership, and (c) in the market development efforts undertaken by individual dealers. Moreover, the association could attack the problem of formulating objective standards for handling fish at sea and changing such currently employed, undesirable methods as "pitchforking."

The opportunities are great for the association to develop coöperative research and marketing projects. The United States Secretaries of State and Commerce and the United States Tariff Commission, after careful review of the industry's import problem, recommended in 1949 that the Congress "provide funds for the appropriate governmental agencies to coöperate with and aid the industry in developing and expanding programs for the further improvement of techniques and facilities for catching, storing, processing, transporting, and marketing of fish.[54] There is room for a legitimate

[53] See Chapter VI.
[54] Cf. "Effect on the Domestic Fishing Industry of Increasing Imports of

increase of federal developmental spending in these areas. In 1945, the government spent about $7.00 for every ton of food produced in agriculture. The figure for salt water commercial fisheries was only 82¢.[55] The federal government has underwritten already, at the industry's request, a $130,000 research effort to determine the feasibility of freezing whole fish at sea for later thawing, processing, and refreezing ashore. The objective of this development is to overcome the industry's fish scarcity handicap. It will be discussed fully in the next chapter, but the move in itself presents a perfect instance of the type of concrete, progressive project for which an alert industry association might solicit aid. Furthermore, a representative from Massachusetts has asked congressional passage of a bill to permit the Secretary of the Interior to spend up to five mlilion dollars a year to encourage the distribution of fishery products.[56] The industry might gain such assistance if it had an association to submit definite proposals which would justify favorable congressional action.

In addition to federal sources, the industry properly might call upon the states of Massachusetts and Maine for assistance. Both states have a substantial stake in the prosperity of New England's fresh and frozen fishery operations. Maine already has pointed the way. That state spends $35,000 annually on a seafood promotion program.[57] Attractive cook books and display materials carry publicity for Maine's many seafood items throughout the nation. Massachusetts, with the New England region's three largest ports, contributes no funds to promote that state's sector of the industry.[58] Nevertheless, the Massachusetts Governor, spurred by the uncertain future of the state's fishing operations, in December, 1949, ap-

---

Fresh and Frozen Fish," *Report of the Secretaries of State and Commerce and of the United States Tariff Commission in Response to House Resolution No. 174, 81st Congress* (Washington, D.C.: U.S. Government Printing Office), 1949, page 68. As matters now stand, the Fish and Wildlife Service does not even have sufficient funds to operate its research vessel most of the year. Moreover, Service plans to investigate such basic influences on fish supplies as tide, current, and temperature have largely had to stay on the shelf.

[55] Source: U.S. Fish and Wildlife Service.

[56] H.R. 7849, "A Bill to further encourage the distribution of fishery products and for other purposes," submitted by Congressman John F. Kennedy in 1950.

[57] Information from Richard E. Reed, Commissioner of Maine's Sea and Shore Fisheries until 1951.

[58] Information obtained from Mr. Francis W. Sargent, Director of Marine Fisheries, Department of Conservation, Commonwealth of Massachusetts.

pointed a committee with representatives from each of the industry's principal groups to explore possibilities for helping the industry. The committee's report in February, 1950, properly stressed specific federal aids, but also it urged that the Massachusetts Industrial and Development Commission coöperate with the industry to promote more widespread use of groundfish. Despite the Governor's pledge to back this objective, no Massachusetts action has followed. If the industry had possessed an association of the type we advocate, prepared with concrete plans for implementing the endorsed objective, then constructive action might now be underway.

Private coöperative arrangements provide still another avenue for an industry association to pursue. Readily available at Woods Hole, Massachusetts, for example, are nonprofit marine laboratories and a recently founded fisheries engineering institute.[59] Recent research of the former has already demonstrated that it might make a sizeable contribution toward solving the industry's fish scarcity problems.[60]

On the market development side, opportunities exist for joint promotion with allied food industries. Soup, wine, and condiment manufacturing concerns, for instance, might be induced to cofeature particular New England species in their recipe and display advertising.

The proposed developmental moves would forge industrial strength at minimum costs to the industry's principal groups. Each group and its respective members, particularly the individual dealers and vessel owners, would retain complete freedom to compete, both in adopting better business practices and in exploiting new ideas. The provision of a high quality floor through action of the industry would not, for example, deter individual operators from seeking to surpass it with their products. Those willing to make the most of benefits offered

---

[59] Specifically, the organizations are the Woods Hole Oceanographic Institution and the Marine and Fisheries Engineering Research Institute, Inc. The latter formed in the Spring of 1951 "to fill a gap between the marine research performed by scientific organizations and the practical engineering aspects required by the fishing industry and other industries with marine problems." (Quotation is taken from "Press Release No. 1," Marine and Fisheries Engineering Research Institute, Inc., Hatchville, Massachusetts, April 6, 1951.)

[60] Dr. William C. Schroeder and other scientists of the Oceanographic Institution's staff have discovered large supplies of "king sized" redfish at trawling depths considerable greater than those currently exploited by commercial fishermen but within comparatively close distances from the main New England ports. See "Along Our Coast," the *Boston Traveler*, December 1, 1952.

by the suggested association would enjoy the greater success. At the same time, however, the policies of all interests in the industry would acquire a positive orientation.

In assuming the initiative through revised union policies and collective bargaining to start the industry moving on a progressive basis, the Atlantic Fishermen's Union would but be following its legitimate entrepreneurial interest under the profit-sharing wage lay system. It would be making organization count constructively in determining the destiny of "America's Oldest Industry."

The crucial question, however, concerns the willingness of the parties to face facts and reorient their thinking. We emphasize that this does not mean any group need pretend affection for the other where no feeling of kinship exists. It does mean that each organized segment of the industry must reëxamine realistically its positions and policies. Their view must be to replace or modify with carefully devised sets of machinery those arrangements which foster conflict and limit competitive ability. Strictly on a self-interest basis, the industry-wide centrally directed Atlantic Fishermen's Union could provide the impetus.

What is the outlook?

CHAPTER VIII

PROSPECTS FOR PROGRESS

W E H A V E O U T L I N E D the major problems which confront New England's principal fishing ports and we have suggested (a) that the Atlantic Fishermen's Union has every reason to take the lead in a frontal attack on those problems and (b) that collective bargaining with revised union policies can be an effective instrument in bringing about sustained and constructive changes. Therefore we have proffered a program for action.

Before proceeding to discuss the outlook for the future, we wish to reëmphasize that when we suggest that the Atlantic Fishermen's Union revise its policies and push for forward-looking programs, we do so because our analysis indicates such a strategy to be in the best interest of the union and its membership. That way lies real protection for the fishermen's incomes and jobs and real security for the union. We do not argue, for example, that the union agree not to use "limits" because we are unsympathetic to the fishermen's normal desire for good incomes; rather we do so because we are concerned about the future of the fishermen's incomes. We realize that our prescription is not an easy one. In effect we are calling for far more from the Fishermen's Union than the vast majority of unions in the United States contribute to other industries. The basic policy of the latter, according to one authority, "has been to drive hard bargains and thereby to get for workers as big a share as possible in the growing output of industry. Unions are not interested in sharing in management or in coöperating with management." [1]

The fact seems plain, however, that under the lay system the

---

[1] Cf. Professor Sumner H. Slichter's discussion of the philosophy and basic policies of American trade unions in "Current Labor Trends and Labor Problems," address before the American Gas Association, Atlantic City, N.J., October 27, 1952, page 7. Professor Slichter carefully points out, however, that there are a few notable exceptions to this statement.

Fishermen's Union must in its own interest feel broad responsibility for the prosperity of the industry in which its members work. It is equally apparent that the union has recognized this for some time: witness the leading role it has assumed in presenting the industry's tariff demands.[2] The difficulty, then, is not that the union fails to appreciate the nature of its role in industry affairs, but rather that to achieve the results it seeks it should change its approach.

Within the foreseeable future what will the union do and what will be the reactions of the other principal groups? Are there forces at work which may tend to pull together the diverse interest groups of the industry, which may render easier the transition from conflict to coöperation, which will underscore the worth of adopting positive policies? Are there real prospects for progress?

The Improved Setting

In the beginning of 1953, several factors supported a show of optimism. The promise of technological advances, propitious changes in the area of resource conservation, and the enlargement of positive governmental assistance provided some basis for favorable expectations.

NEW PROCESSES AND PRODUCTS

The most hopeful technological development in the industry in the last generation — the freezing of fish at sea — may be about ready to bear fruit. Indeed, that is why we have deferred discussion of the process until now, even though the roots of the change go back to about 1948.[3] Not only is the freezing of fish at sea promising because it may well prove as revolutionary a stimulant as the introduction of the filleting process did after 1921, but, equally important, it has had the full support of all major interest groups, particularly the Fishermen's Union. In a sense, the project thus far has involved the nearest thing to a coördinated research effort in the industry's long history. Also, it presents a picture of governmental assistance to the industry at its best.

[2] See Chapter VI.

[3] For discussion of the historical aspects of this process see J. M. Lemon, and C. B. Carlson, "Freezing Fish at Sea," *Fishery Leaflet 278* (U.S. Fish and Wildlife Service, Washington, D.C.), January, 1948.

The apparent scarcity of key species of fish in New England waters provided the initial impetus for action. Technologists of the U.S. Fish and Wildlife Service, noting the increased cost and difficulty of obtaining "payloads," started experiments in 1948 to determine the feasibility of freezing fish at sea for later thawing, processing, and refreezing ashore. The aim was to extend the range of fishing operations while maintaining or if possible improving the quality of fish landed. The technologists adopted the objective of freezing for further processing ashore rather than developing a complete "factory" ship wtih facilities for finalizing the product at sea because the former might be adapted for use on present fishing craft at minimum cost.[4] Moreover, the former method seemed likely, if successful, to cause less disturbance to the operating structure of the industry and to avoid labor recruiting and operating difficulties posed principally by the rigors of actually processing fish in the storm-afflicted northwest Atlantic. Also, the technologists knew that the West Coast tuna fishery had employed somewhat similar techniques successfully for years. The wisdom of taking this more limited objective received some confirmation by the end of 1951. A full-fledged factory ship financed by private interests in 1950 succumbed in the face of the obstacles we have cited.[5]

The Fish and Wildlife scientists reported the results of their initial studies in May, 1949. These showed that fillets of fish frozen at sea were of better general quality than fillets of fish iced at sea (the kind which currently dominates the industry's output).[6]

---

[4] For a discussion of factory-ship experience in the United States see A. W. Anderson, "Technological Development in Fisheries with Special Reference to the Factory Ship in the United States," United Nations Scientific Conference on the Conservation and Utilization of Resources, U.N.E.S.C.O., April 12, 1949. Mr. Anderson has headed the Fish and Wildlife Service's Branch of Commercial Fisheries and is on leave to perform similar duties in the U.S. Defense Fisheries Administration.

[5] For a full descripiton of this venture see "Packaged Fish in Novel Haul of Boston Trawler," *New York Herald Tribune*, July 24, 1950. The vessel, the *Ocean Life*, operated briefly from Rockland, Maine because the unionized fishermen and shoreworkers in the Massachusetts ports would have nothing to do with it. It was the inability to surmount technical obstacles as well as union opposition, which really sounded the death knell for the experiment. The vessel has reconverted to conventional redfishing out of Gloucester.

[6] Cf. "Research in Service Laboratories," *Commercial Fisheries Review*, July, 1949, page 17.

The findings sparked definite interest in industrial circles, and the sequence of events which followed is worth noting briefly. We have already mentioned the conference of representatives from all sectors of the industry convened by the governor of Massachusetts in December, 1949.[7] That committee made several recommendations,[8] but the one it really got behind was the freezing of fish at sea. With the Fishermen's Union and vessel owners in the lead, industry interests succeeded in convincing Congress to appropriate about $130,000 for full-fledged trials. The trawler *Delaware*, one of twelve Boston vessels bought by the federal government in 1949 and leased to German operators, was returned to that port in late 1950 and revamped for the project at a cost of about $40,000.[9]

After extensive tests, Joseph F. Puncochar, Chief of North Atlantic Technological Research for the Fish and Wildlife Service, reported that the process soon might be about ready for commercial adoption.[10] Not only had the process been tested experimentally and the "bugs" largely worked out, but also the *Delaware* had operated commercially and sold the product to a few leading dealers through the New England Fish Exchange. The dealers found processing results as good as the technologists had forecast. In handling the brine-frozen fish, the buyers used successfully the tested sequences which the technologists had worked out.[11]

[7] See Chapter VII.

[8] Other major recommendations included: Immediate implementation of the Northwest Atlantic Convention for conservation; further Fish and Wildlife investigation of depletion problems; Reconstruction Finance Corp. loans to improve the industry's facilities; increased fish purchases by the U.S. Armed Forces; and establishment of a coöperative program with the Massachusetts Development and Industrial Commission (a state agency to promote Massachusetts industry) to increase fish sales. See "Urge R.F.C. Loans to Modernize Fishing Fleet," the *Boston Globe*, February 27, 1950.

[9] Cost information obtained from Mr. Joseph F. Puncochar of the U.S. Fish and Wildlife Service, who directed the research on freezing fish at sea.

[10] Mr. Puncochar graciously reviewed the process and its prospects in an interview with the author on October 23, 1952.

[11] The following series of articles in the *Commercial Fisheries Review* discuss the process fully: J. C. Hartshorne and J. F. Punochar, "Freezing Fish at Sea — New England, Part I, Preliminary Experiments," *C.F.R.*, February 1952, pp. 1–7; H. W. Magnusson, S. R. Pottinger, and J. C. Hartshorne, "Part II, Experimental Procedures and Equipment," pp. 8–15; C. Butler, J. F. Puncochar, and B. O. Knake, "Part III, The Experimental Trawler *Delaware* and Shore Facilities," pp. 16–25; and C. Butler and H. W. Magnusson, "Part IV, Commercial Processing of Brine Frozen Fish," pp. 26–29.

The most gratifying thing about these remarkable developments in less than three years time is not the splendid technical performance of federal fishery researchists, though it was indeed superlative. More important is the fact that fish executives, particularly in Boston, and the leadership of the Fishermen's Union coöperated and contributed what was useful from their own experience. Thus the promising new technology is not something to which the industry needs an introduction. This is significant because the next steps are up to them.

What will the inauguration of freezing fish at sea mean to the New England industry? Obviously, it will allow a much wider range of fishing operations. This will alleviate somewhat the fish scarcity problem by spreading fishing effort over a larger area, even to the enormous and not too fully exploited Grand Bank of Newfoundland.[12]

The process may mean lower costs per unit of catch by significantly increasing "payloads" per trip. Where vessels, especially from Boston, have brought home from one-third to one-half the load of which they were capable in recent years,[13] they may operate near capacity with the freezing apparatus, and they can also be sure of landing the highest quality fish. All fish taken up to the last day or two of fishing will be frozen at the peak of quality as they come from the ocean; the end of trip haul will be iced down for the prime fresh fish market and will probably spend but half the time in transit that most of the catch must now.

There are other less obvious but important benefits which may be expected. For one thing, the system will narrow sharply the area for dispute over quality. There may still be some "No. 2" fish, but most of the catch clearly will be either prime fresh or frozen fish. Hence the process will implement the success of the sellover reforms we have suggested. Secondly, the process will cut dealers' costs and better their marketing situation. Experience already available from commercial processors shows that the frozen fish have a higher fillet yield, for they are firmer and easier to cut. Moreover, such new products as cross sectional fish steaks and whole fish prepared especially

<hr>

[12] At present only Gloucester redfish vessels make more than an occasional trip to the Grand Bank.

[13] This has been true even where no "limits" have applied.

for baking or broiling may enable the dealers to broaden their market. Both of these moves would tremendously boost processing yield. It goes almost without saying that the new techniques will benefit by-products manufacturers because waste formerly discarded at sea will become available.[14]

Freezing at sea promises also to act as a stabilizing influence in the industry. Processors will be able to level out their operations, thereby steadying shoreworker-employment and avoiding inefficiencies and costs stemming from uneven use of overhead capacity.[15] This will follow because storage will be possible one stage earlier than at present in the production stream — the dealers will be able to buy for future processing. Moreover, this factor in turn will make for more price stability in the primary market. During the spring and summer months when fishing conditions are best and supplies most abundant on the banks, the dealers will be able to offer a stronger demand. Under such circumstances, it will be necessary that the union make sure that landings are large enough to maximize earnings as the increased demand will make possible greater returns for more production.[16] Unless there were to be a significant increase in the size of the fleet, "limits" would almost certainly be out of order and not in the interest of either the fishermen or the other principal groups.

An important drawback of the freezing at sea development is that initially it will require substantial investment by the vessel owners. At present prices, the new equipment and alterations would probably cost between $30,000 and $40,000 on large trawlers of the type sailing from Boston.[17] The owners are conservative business men as a whole, so that this may delay the conversion to new methods. The union might be able to speed the change, however, by giving the owners some assurance that in the good fishing months the fishermen

[14] Under present methods, the fishermen head and "gut" (remove the entrails) the larger groundfish (haddock, cod, etc.) at sea.

[15] One of Boston's more progressive firms, Fulham Bros., noted this in 1952 from the experience the concern had with frozen fish bought from the *Delaware*, according to Mr. Puncochar of the Fish and Wildlife.

[16] That is, the demand will shift to the right during the months of most favorable fishing.

[17] $40,000 was the cost on the *Delaware*. That experience might make current costs less.

will not hold production back. In view of the fact that demand should be better under the system in those months, such assurances should be possible.[18]

There is one other obstacle to note. Successful inauguration of the freezing at sea will require some investment on the dealers' part.[19] They will have to install equipment for handling the frozen product, including means for quick removal of the fish from the pier to freezers. No longer will it be possible to let the fish lie around in an ordinary cart while they await processing! In a very real sense, the adoption of the new method will require coöperation of dealers and vessel owners, for a sufficient number of both must acquire the necessary equipment or the process will benefit no one. It is perhaps fortunate that the Boston industry has had the closest association with the new technology and that a Boston-type vessel has served as the "guinea pig," for the presence of the dealer-owner combination type of operation in that port should at least make the need of dealer-owner coöperation much less of a hurdle. Once Boston had swung somewhat fully and successfully into the new system, it might share its experience in implementing a similar conversion in the other ports. The job required might be carried out quite well by the type of association we have suggested in Chapter VII.[20]

Before we pass on to discuss other developments, we should note that over and above the merits we have listed, the freezing at sea process may improve demand simply because it seems so likely not only to raise the level of quality but to insure uniformity of that high quality as well. In summary, then, the new technology promises lower costs, better quality, and improved demand, and a narrowing of the area for quality disputes. Even so, it presents the major groups with a sharp challenge to follow through with the kind of progressive policies which can open up a new era. We shall discuss later in this chapter the prospects for such a turn of events.

Other developments invite further favorable expectations. One is the recent introduction of new processing techniques. During 1951 leading dealers began breading, precooking (in some instances deep

---

[18] We assume, of course, that other factors, e.g., meat competition, do not cut demand.

[19] The amount of money involved we do not know. However, indications are that it would not be very substantial.

[20] See Chapter VII.

frying) and freezing fish for one-pound packages.[21] While no data are available on the sales of these lines, there is some evidence that they have enjoyed growing popularity. The largest processor at Boston Fish Pier, for example, broke the pattern of virtually no second shift operations in the fall of 1951 in order to meet the demand for these products.[22]

The importance of this development cannot be overemphasized, for it strikes directly at one of the industry's key marketing problems, the housewife's difficulties with fish preparation. Not only does it ease preparation, but it insures uniformly high quality. Hence this innovation complements nicely the coming freezing of fish at sea, for it attacks foursquare the industry's market problem. It may, too, prove the answer to increasing exploitation of the lesser known and appreciated species in New England waters. Pollock, for example, might well turn out to be a big seller as a precooked item.

The introduction of precooking techniques promises to have one other favorable effect. The leading dealers, having invested substantially in promoting a good quality product, are increasingly becoming aware of the need for the industry-wide quality standards and exchange of technical knowledge which we suggested in the previous chapter.[23] They know that one processor with poor technique and little in the way of quality control can do unlimited damage to demand.

In the light of the above considerations, a move took place in the fall of 1952 which might well be the forerunner of the type of industry research arm the industry needs. The fishery technicians of New England, consisting mainly of men from the major processing firms and state and federal governments, formed an organization called the New England Fishery Technologists.[24] Their aims include coördinating and raising standards of quality.

[21] Even before precooking started, the industry (or at least the more progressive dealers) had begun to market quick frozen one-pound fillet packs; they too have enjoyed good sales.

[22] The firm, the A & P, could not obtain sufficient friolator equipment to meet demand with day-shift operations only.

[23] See Chapter VII.

[24] Cf. "Fishery Technologists Organize N.E. Group," the *Boston Herald*, October 19, 1952. Mr. Joseph F. Puncochar was chosen as the first president of the organization.

All the propitious changes in the frozen food area within the industry mesh perfectly with the tremendous growth in frozen food markets in recent years. Since 1939, frozen groundfish fillet sales have nearly tripled [25] and the growing availability of facilities for displaying and selling frozen foods at retail augur well for much further expansion.[26]

RESOURCE PROTECTION

In June 1952, U.S. Fish and Wildlife biologists accompanied the crew of the Boston trawler *Michigan* to sea on two regular trips to test the use of larger mesh nets as a device for preventing the premature capture and waste of undersized fish, especially haddock. The fishermen, members of the Atlantic Fishermen's Union, coöperated unstintingly in the experiments. The tests demonstrated (a) that the larger mesh allows at least 27 per cent — the fish which would ordinarily be caught prematurely — to escape unharmed and (b) that eliminating such fish from the catch does not cut the size of the saleable haul. The *Michigan* was "highliner" (landing more fish than any other vessel for the day) at Boston Fish Pier after both trips.[27]

In Chapter VI we stated that developments between March 20, 1951, when the commissioners held their first meeting with the advisory group, and the end of 1952 augured well both for the future of the Northwest Atlantic Fisheries Convention and the New England fishing industry.[28] The success of the *Michigan* experiments culminated over sixteen months of close coöperation of all segments

[25] Consumption of imported and domestically produced fillets amounted to 192,178,000 pounds in 1951. (Data were obtained from the U.S. Tariff Commission.)

[29] Some 65 per cent of all food stores in the 55 cities from which the Bureau of Labor Statistics collects price data have one or more frozen foods cabinets. See "A Survey of Frozen Food Cabinets and Prepackaged Fresh Meat in Retail Food Stores of 55 Large Cities," U.S. Department of Labor, Bureau of Labor Statistics, December, 1951.

[27] The beneficial results surprised even the fishermen, who have generally supported the proposal for mesh net regulation for many years. For a description of the *Michigan* experiments see J. R. Clark, "Experiments on the Escape of Undersized Haddock through Otter Trawls," *Commercial Fisheries Review*, September, 1952, pp. 1–7. The larger mesh caught 4384 fish; 1190 escaped — and virtually all were immature.

[28] See section on Haddock.

of the industry with the Convention Commissioners. As a result, they planned to install a larger mesh net regulation by spring, 1953. If the optimistic estimates of some officials work out, the regulation will increase the New England haddock catch by as much as 35 per cent in 15 years. Whether or not the use of savings gear proves that successful, the convention thus far has served as a unifying force in the industry. As such it has reinforced the catalytic action of development of freezing fish at sea.

STATE AND FEDERAL ASSISTANCE

The record reviewed in the foregoing sections tells volumes about the proper role for government in the New England fishing industry. But there is still more to the story.

At the state level we have noted that Maine has been exceptionally active, whereas until the end of 1949 Massachusetts had taken little or no interest.[29] The situation in that state is changing, however. An important reason for this is that the young dynamic Director of the state's Marine Fisheries emerged from the committee action of 1949 and early 1950 with the complete confidence and respect of the industry's leaders.[30] The latter have given him full support as one of the three commissioners administering the International Convention for the Northwest Atlantic Fisheries. This fact takes on added significance when one recalls the bitterness which surrounded the whole issue of selection of the commissioners when the enabling legislation for the convention was before Congress.[31]

The Director has pushed hard to obtain funds from the Massachusetts legislature for promoting the products of the fishing industry. Despite a series of setbacks, an idea which he sponsored in 1952 is receiving a somewhat favorable reception. The proposition

[29] That is, Massachusetts had taken no interest for some twelve or more years. In 1938 the state coöperated with the federal government and the city of Gloucester to furnish a new fish pier for that port. And of course in 1914 the state provided the Boston Fish Pier on favorable terms.

[30] The Director is Francis W. Sargent. A graduate of Massachusetts Institute of Technology, Sargent spent the first year or so of his appointment getting acquainted with the people and problems. He accompanied a crew on a regular trip to sea.

[31] See Chapter VI.

is to start a state fisheries college which would furnish the industry with men equipped with technical "know how." To bolster his argument, he has pointed out that the only institution of this kind in the United States is one located at the University of Washington on the west coast. On the other hand, New England's principal competitor — Canada — has such a school in St. John, Newfoundland.

The proposal is not far-fetched as some might claim. A fisheries degree might well be added to those available from the University of Massachusetts. In view of the rather haphazard role of technology in the industry's affairs in the last thirty years, it seems reasonable to conclude that present conditions would be far better if a similar proposal made in 1920 had received sufficient backing to materialize.[32] Also, the available evidence indicates that research and technology will play no mean role in the industry's future.

At the federal level, government agencies conducting consumer education programs have begun introducing the industry's products through television.[33] The Fish and Wildlife Service has had fishery educational motion pictures, in which the New England industry has shared on over 95 per cent of the nation's stations. Its fishery marketing specialists and home economists have appeared in person on a number of programs to teach consumers how and why fish should be used. Finally, the U.S. Department of Agriculture has given fish and shellfish considerable time on its TV programs, both network and local, which are designed to call the public's attention to foods on the department's "Plentiful Food List." These activities point the way to improved consumer demand.

All the developments reviewed in this section indicate an improved setting for the industry. Technological change promises weapons for the fight to subdue supply and quality problems and to capture stronger markets. Present and prospective government assistance is aimed at achieving the same goals. Will the industry's major economic groups, particularly the Fishermen's Union and the dealers, mend their differences and make the most of their opportunities to move forward together?

[32] Cf. "Urge College of Fisheries at Harvard," the *Fishing Gazette*, October, 1920, page 11. Mrs. Kathryn Maloney of Gloucester is the prime mover for the founding of the fish college program.

[33] Cf. "Fish on Television," *Commercial Fisheries Review*, July, 1952, page 34.

The Outlook for Coöperation

The analysis developed in Chapter V makes it clear that evolving workable relationships for solving their common problems constitutes no mean task for the dealers and the Atlantic Fishermen's Union. They must forget the past — not an easy thing to do when the memories are so bitter. More important, they must dispose of the institutional features of the primary markets in each port which intensify their differences — indeed which practically provide them with a "built-in war." Constructive changes can be accomplished and better market arrangements achieved if the parties will but couch their thinking in more positive terms. Is that likely in the near future?

We suggested in Chapter VII that the best opportunity for straightening matters out exists in Boston because in that port there is a collective bargaining relationship between the dealers and fishermen.[34] Are the prospects good for bargaining for the kind of institutional changes suggested in the previous chapter?

On the surface one might think that the outlook was promising. After the Korean episode the parties "mellowed" somewhat as catch values increased and the "sellover" became less frequent. The value of landings in the last half of 1950 exceeded that for the same period the year before by about a million dollars, or nearly 20 per cent. For the year 1951 they reached $14.3 million, surpassing 1950 by about 6 per cent. This situation reflected in part perhaps the fact that the industry was not subject to price controls and, under the lay, the fishermen shared in more favorable prices.[35] Moreover, both the union and the dealer-owners have been coöperating since 1950 with government officials on both the freezing at sea project and conservation difficulties. All this seemed to reflect a change in attitude. Finally, late in 1952, the secretary of the employers' association indicated that before too long the dealer-owners, with perhaps one significant exception,[36] will be willing to negotiate a new agreement

[34] See Chapter V also.

[35] At the outset in January, 1951, fresh fish and seafood were exempted at all levels of production and distribution under the general ceiling order. Frozen fishery products were covered until May 17, 1951, when the ceiling was lifted on those items. The complaints of the industry because of the discrimination between fresh and frozen fish for purposes of stabilization policy helped in no small measure to bring the latter action.

[36] The exception is the R. O'Brien Co. See Chapter IV, page 54.

with the union and to include on the agenda consideration of the "sellover." He pointed out in an interview that for the most part the employers are genuinely anxious to forget the bitter past; he is working hard to bring them to negotiate. As evidence of their good will, he indicated that without a formal new agreement the owners have increased "broker" payments from $5 to $7 per day and the maximum from $50 to $70 per trip.[37]

On the union's side, Captain McHugh showed willingness to negotiate, though of course he had to speak for himself and not necessarily for the union. He agreed that having a competent, impartial party paid jointly and well by the dealers and union might be of notable assistance in straightening out the persistent price difficulties on Boston Fish Pier.

But the picture is not so bright as it appears. The most important reason why the dealer-owners are beginning to think in terms of securing a new contract is because without one they are at a disadvantage in day-to-day disputes over such things as "broker pay" — wages lost on a breakdown at sea. The fishermen may arbitrate if they feel like doing so, but in the absence of a binding grievance machinery clause, they have been free to threaten and carry out a stoppage sufficient to win their point. The owners are thinking, too, in terms of gaining whatever advantages the Taft-Hartley Act has to offer, particularly the protection of the provision that allows legal action for noncompliance with a contract. So far as taking action to revise the market mechanism in the port, however, most of the employers are quite unenthusiastic. They point to the fact that the federal court approved the rules and there is nothing wrong with them.[38]

The approach of the employers would be a more promising one if it were coupled with a genuine desire to straighten out the "sellover" difficulty. This is the nub of the problem in Boston. As we pointed out earlier, McHugh might go along personally with the kind of

[37] For explanation of the term "broker" see Chapter V.

[38] The dealer-vessel owners' association executive secretary, Mr. Thomas D. Rice, pointed out that the system of impartial arbiter determination of quality would break down because the latter would have to find against the fishermen too often. This may be so, but as we pointed out in Chapter VII, even this might not be too serious a difficulty if the fishermen were sure that the only factor at work in the "sellover" were quality.

arrangement we suggested in Chapter VII, but it is probable that he more reliably reflected membership sentiment when he voiced strong reservations about the effectiveness of such a change. His objection, in brief, amounted to a deep-seated doubt that the dealers would manifest the good faith necessary to go along with such a system, even if agreement were reached to try it. At the same time, McHugh is probably completely correct when he says that there will not be another Boston contract unless it clears up the "sellover" difficulty; and basically neither his thinking nor that of the membership has changed from the long-held belief that the device must be wiped out.

Perhaps if the practice of freezing fish at sea is adopted, it will expose the present "sellover" setup for what it is, a grotesque mechanism; and possibly it might render the sellover unnecessary by providing two clear-cut kinds of fish for sale in the first sales market. Barring such a turn of events, however, and recognizing that even with that device in force there may still be claims that the "sellover" applies, it would seem that only a reëxamination of the New England Fish Exchange rules by federal authorities with a view to installing the changes that must be made, will bring about the necessary order. Perhaps such action should be taken in view of the court sanction given to those rules. It would be better, of course, if the parties could work out their own solution, for that would reflect precisely the kind of change in attitudes the port so desperately needs. At any rate, until the matter is settled, the basic structure of conflict will tend to dominate except in those periods when a strong demand for fish obscures the problem. And the latter is not likely on any sustained basis, in view of the competitive pressures, until the parties take steps of the kind we have advocated elsewhere. Unfortunately, such steps would in turn wait upon the development of just the kind of relationships which the present market mechanism tends to make impossible.

Turning to the other ports, the outlook is not much more propitious. Things remain about the same in New Bedford.[39] When demand is high, as it was in the latter half of 1950 and through 1951, there is little difficulty. Let conditions revert to their 1949 status, however, and the skirmishes over quality and weights will renew in ear-

[39] As of 1952. See the Postscript to this book.

nest. At Gloucester, the prospects are a bit more favorable. The area of difference over the "percentage" has narrowed. The dealers are cutting fillets from small fish they used to argue were "trash," because redfish have become increasingly less plentiful. Everyone in the port knows this. Consequently, the dealers' chances to claim that a large portion of each trip is trash have diminished. While this has lessened frictions between the parties, one other factor deserves attention. In 1952 the union stated that the dealers were considering the possibility of having a winter and a summer price for redfish similar to the arrangement that prevailed under Office of Price Administration ceilings during the second World War.[40] Such an objective, of course, has long been close to the union's heart. As such, the mere willingness of the dealers to think about it makes for better relations with the union. It would have little chance of success, however, without adequate demand; in fact, it might make matters worse if demand flagged after the dealers had committed themselves and they turned to other devices to recoup their position. Nonetheless, the situation is one in which, with a fairly good break in demand conditions, the parties might get started on the right foot and take those steps necessary to bulwark demand in time to make the system work. In short, if the system did not lead to disillusion, it might serve to bring the parties together to work out their common problems.

Obstacles within the Fishermen's Union

The character of the Atlantic Fishermen's Union presents a definite hazard for the prophet of optimism. The membership is highly individualistic, "crisis-minded," and very short-run in its thinking.[41] In the absence of a problem both imposing and immediate, the membership is prone to slide along even without meetings. During 1952 the Boston policy-making local went from spring to fall without a meeting.[42] The continued cavalier attitude on the problem of prices of supplies paid for from the fishermen's share of the lay forms per-

---

[40] The Office of Price Administration fixed redfish ceilings "ex-vessel" at 3.75¢ per pound for the April–September period and at 4.25¢ for the October–March period annually. See *Fishery Market News Supplement* (U.S. Fish and Wildlife Service, Washington, D.C.), March, 1944.

[41] See Chapter IV.

[42] Information obtained from Captain McHugh on September 19, 1952.

haps the best illustration of membership inertia. Since the earliest days of the union, the men have charged that vessel owners deduct from the fishermen's share of the lay the cost of food and oil computed at retail prices. Since the Boston fleet, for example, buys more than $750,000 worth of such items annually, the fishermen insist that the owners must be getting a "kickback." (The latter, of course, vehemently deny this.) Pleas of the union's leaders to move the men to form a coöperative for purchasing supplies, or at least to insist that orders be filled on a contractual basis annually after competitive bidding, have fallen more or less on deaf ears. Although a constant source of irritation, the matter simply has not pressed hard enough on the men to incite them to take action.

Unfortunately, the current leadership structure of the union tends to perpetuate such a state of affairs. As we have noted in Chapter IV,[43] the burdens of administration are for the better part concentrated upon Captain McHugh. He has perhaps done too good a job; he expends more than normal effort for one man and that satisfies the membership. It goes without saying that much of this effort must be concerned with details; there would be little possibility of McHugh's undertaking the beginnings of a coöperative purchasing venture, for example, even if the men were to authorize the necessary related expenditures. On the latter score, moreover, it should be stressed that the membership is most reluctant to open the purse strings, even for McHugh.

The problem of membership inertia and shortsightedness is not an insoluble one, even though it is one that will take some time to meet. A good start on eliminating it could be made by broadening the union's leadership structure. An executive board, composed partly of members at large and partly of men chosen to represent each port and not exceeding seven in number, would be of great assistance. It would spread responsibility, afford an opportunity for more members to come face to face with important problems affecting the entire union, and it would turn up potential officer material for the organization.[44] It would bring the rather far-flung rank and file into closer communication with over-all policy problems. It would not have to be financially burdensome; six of the members might be paid only

---

[43] See section on Atlantic Fishermen's Union.

[44] As we have noted, the Union has had some very poor minor officials.

when actually engaged in union business; the secretary-treasurer, who is a full-time officer, might serve as the seventh member and chairman. The arrangement would make possible real delegation of power to make decisions on matters of the kind involved in the program offered in the previous chapter, but it would avoid overcentralization of authority. Until steps of this kind are actually taken, the prospects for membership support of a positive program are quite bleak.

Interunion Difficulties

The course of action we have set forth in Chapter VII makes provision for eventual participation of all the major groups in the industry, including the Seafood Workers' Union. We have pointed out that through the years the Fishermen's Union and the Seafood Workers' have not come to work together, partly because in a sense the latter almost have more in common with the dealers than with the fishermen, so far as the shoreworkers' position in the market structure is concerned.[45] Now relations between the organizations promise to undergo further strain because of jurisdictional differences in Maine.

Although the Fishermen's Union has had indifferent success in organizing the fishermen and securing contracts in Portland and Rockland,[46] it has organized the largest filleting plant in the latter port and is seeking to represent the shoreworkers at one plant in Portland.

These moves to recruit fillet workers represent a new turn in union policy. The union has jurisdiction over such workers under its charter from the Seafarers' International, AFL. In exercising it, however, a direct clash results with the Boston Seafood Workers' Union, International Longshoremen's Association. The latter has for some time considered the Maine ports, particularly Portland, to be its job territory. In 1952, the Seafood Workers had contracts with six Portland firms, but had not achieved any in Rockland.

The action of the Fishermen's Union in attempting to organize

---

[45] See Chapter IV.

[46] As this is written at the end of 1952, the union has but few members in Portland and no contracts. It claims, however, to have a high percentage of the Rockland fishermen organized, but still no contracts in that port. See, however, comments in the Postscript to this book.

shoreworkers in Portland and Rockland may be a policy aberration, not a new line of organization. The difficulty is that the tensions set up by such a policy may seriously affect interunion relations in the other principal ports. The breach was not serious as 1952 drew to a close, though it was bad enough to have some repercussions. Looked at from a longer run standpoint, therefore, the policy is inherently dangerous. Unless it is replaced by some provision for a joint effort by the two unions, it threatens to become a barrier that would forestall any coöperation for mutual objectives.

The Tariff Question

The position of the major groups in the industry on the tariff question constitutes the final check-rein on our optimism.

The substance of the situation may be put very briefly. In Chapter VI we noted that the United States Tariff Commission refused late in 1952 to recommend greater tariff protection for the industry. The latter has appealed that decision; they feel it was unjustified and possibly "political." [47]

The industry correctly points out that it would secure a substantial addition to the tariff wall if the import authorities were to consider as a basis only the fact that the industry is entitled to as much protection as it enjoyed before the second World War.[48] In connection with the petition for rehearing, the New England interests have worked out a compromise proposal with other members of the National Fisheries association who vociferously opposed the New Englanders during the original proceeding.[49] Essentially, the plan would raise the basic tariff figure from $1\frac{7}{8}\cent$ to a flat $2\frac{1}{2}\cent$ per pound and would make provision for a flexible quota that would allow importers about 37 per cent of the American market, on an annual basis. Roughly, this is the percentage which imports bore to domestic consumption in 1951. The plan as proposed would be operative for five years and then subject to upward or downward revision as U.S. consumption trends dictated. The Wholesaler members of the NFI apparently

[47] Mr. Rice of the Massachusetts Fisheries Association has openly made this charge. In part, then, the hope of the parties in the petition for a rehearing was that the change of administration in the fall of 1952 would favor their case.

[48] See Chapter VI.

[49] See Chapter VI.

agreed to such a proposal in exchange for the New Englanders' promise that they would join in "an advertising and publicity plan financed by an assessment of so much per pound of domestic production and of imported fillets." [50]

While the latter constructive aim of expanding the market is all to the good, the diehard tariff attitude of the parties is a bit foreboding. It shows how firm a hold thinking in restrictionist terms has in the industry. One can sympathize with the industry's attempt to protect itself, but the difficulty is that every push by the parties for tariff aid involves tremendous effort — effort which might more profitably be dedicated to formulating and prosecuting positive programs. Perhaps worse still, such strenuous tariff efforts tend to divert the parties' attention from the fact that sooner or later they must concentrate on their real difficulties.

[50] The compromise also spelled out the basis for sharing of quotas among foreign nations. This information is contained in "Essential Points of Proposal for Limiting Imports," a copy of the memorandum of agreement kindly furnished by Mr. Rice of the Massachusetts Fisheries Association on November 8, 1952.

POSTSCRIPT

THE DEVELOPMENTS we have described take the industry up to 1952.
The lines that follow attempt to summarize the rapidly moving events
between that time and early 1954.

On the gloomy side, catch values declined in 1953, and the industry once again asked tariff relief. (The plea received less support
than in the past, however, even from New England interests.) The
Tariff Commission's decision is still pending, but it is worth noting
that quite favorable meat supplies and prices (39¢ per pound ham-
burger) probably hurt the industry more than foreign fish. Indeed,
imports actually declined between 1952 and 1953 by about 17 million pounds, or 16 per cent.

Among the major ports, New Bedford's situation seemed particularly bad. Only about seven fillet houses — with three working full
time — were operating in the Fall of 1953. The port seemed hardest
hit, too, by another problem — high personal injury and indemnity
insurance rates for fishermen. Despite much discussion in New Bedford and the other Massachusetts ports affected, the vessel owners
and the Fishermen's Union seemed to make little headway in attack-
ing the real problem: how to cut accident rates. New Bedford lost
some 42 vessels to the Florida shrimp fishery because of very low
vessel earnings. This development and the insurance cost problem
helped increase the friction between the shoreworkers' unions and the
Fishermen's Union. The former claimed that the Fishermen's Union
was not taking a sufficient interest in preserving the port's fleet.[1]

A series of labor disputes immobilized Gloucester for an aggregate
of better than two months in 1953. First, the fishermen walked out

[1] Relations with the New Bedford owners also deteriorated. The Fishermen's
Union charged the owners with wholesale violation of contractual union security
arrangements. This, of course, tended to dim greatly, for the time being at least,
any prospects for real coöperation on the parties' joint problems.

over the weight problem with the dealers, but nothing was really settled. Then, the seafood workers struck for better wages. Finally, the Fishermen's Union tied up again for a month before a new agreement (very similar to the 1952 New Bedford contract) was reached with the owners. The latter dispute brought to the fore a new boat owners' association which promised to be more active than its older competitor. The new association implemented plans for the purchase of a marine railway, and by the end of the year was well on the way to promoting coöperative arrangements for reducing its members' vessel service costs.

At Rockland, Maine, the Fishermen's Union won a recognition agreement on General Foods' vessels. After an extended dispute, the parties agreed further to operate one redfish trawler on a new wage plan at the request of the company. Under the plan the lay is scrapped. All hands receive a guarantee per trip with a bonus for landings of redfish above a trip standard. The guarantees differ by job classification and compare favorably with the 1951 earnings we have listed in Chapter V. Furthermore, those covered by the plan receive other company "fringe" benefits as regular employees. The plan worked well enough up to early 1954 for the company to contemplate its extension to other vessels. General Foods made other news in 1953, too. The company re-acquired about one half dozen of the vessels sold to the U.S. Army in 1949 for use in Germany. The vessels will probably operate out of Rockland and Halifax, Nova Scotia. Indeed, at the year's end the company was seriously considering moving lock, stock, and barrel from Boston to Rockland, a shift which would be a severe blow to Boston.

The situation was not all gloom, however, even for the harder pressed Massachusetts ports. In many ways 1953 was a bright year, at least potentially. The haddock mesh net regulation took effect in June; the limited evidence available suggests that the regulation spares immature fish and may actually enhance the catch of larger, more valuable haddock at the same time. The technologists on the "freezing of fish at sea" project report continued progress despite a sharp setback due to a mysterious fire which severely damaged their experimental vessel. Also, the more progressive owners are trying out new sounding and other electronic devices. Use of such instruments may well enable the exploitation of supplies heretofore undiscovered.

Ashore, many of the technical difficulties which limited the use of filleting or automatic cutting machines, were overcome. Such machines significantly increase productivity and thus far have been taken in stride by the shoreworkers' unions. All in all, progress in technology seems to hold all of the promise for the industry which we indicated in earlier chapters.

The outstanding feature of the period, however, was the continued boom in pre-cooked fish, especially the new "fish sticks" products. Indeed, those concerns with such pre-cooked items were in many cases straining to fill sales schedules with extra shifts. Even in Gloucester where good raw fish are readily available housewives were reported to be buying the pre-cooked items in volume. Properly handled, the pre-cooking development can provide the same stimulus as did filleting 30 years earlier.[2]

Finally, there was on the horizon the prospect of constructive aid from the Federal Government and the Commonwealth of Massachusetts. Under a bill proposed by the Massachusetts senators, and backed by the industry's leaders, the Secretary of the Interior would receive between three and four million dollars a year, from import duties now allocated to the Secretary of Agriculture, to support research and market development programs. Since these funds, when they become available, will be for the entire United States fishing industry, the New Englanders' success in deriving benefits will depend in no small measure on their ability to compose internal differences and unite behind a well conceived set of proposals. Similar cooperation will be needed if market development assistance, currently under study by the Departments of Commerce and Natural Resources in Massachusetts, is to materialize.[3]

[2] As of early 1954 a very large part of the fish used in pre-cooking was imported block-frozen cod. Hence the benefits of pre-cooking may by-pass New England fishermen and vessel owners unless they concentrate on becoming competitive.

[3] Fortunately, as this is written, the industry is drawing up detailed proposals for presentation at Congressional hearings on the proposed federal aid bill.

THEORETICAL SUPPLEMENT

THE PRICE-MAKING PROCESS AND RESTRICTIVE
UNION POLICIES

As WE HAVE EXPLAINED, the Atlantic Fishermen's Union has adopted restrictive policies, including direct catch limitations, where the members of the Union have felt that such moves would protect their incomes. In addition, the union has clashed with the dealers in the main ports over the "proper" prices to be paid for fish landed.

The rationale of the union's action and the resulting conflict with the dealers is explained below in terms of theoretical models of the market mechanism. Before launching the discussion, however, we should point out that we have assumed generally that the demand for fish is elastic.[1] As Chapter V has indicated, fish compete closely with meat and other protein foods for consumer patronage.[2]

THE BASIC MARKET MODEL

Exhibit A shows the various functions and the relations between them. On the horizontal axis are two scales. These show the relationship between pounds of raw fish and pound of fillets which are cut from the raw fish. For purposes of this exposition, we are assuming a 40 per cent yield.[3] Thus, when the dealers demand 25 pounds of raw fish, they are equivalently demanding 10 pounds of fillets.

On the vertical axis we have measured both the price of raw fish and the price of fillets. ZR is the demand curve for fillets presented to the main port dealers by wholesalers and others. This market de-

---

[1] The curves in the diagrams are presented as straight lines for convenience only.

[2] See Chapter VI.

[3] This is approximately the yield which the industry derives in filleting haddock.

mand curve is a composite of the somewhat individualized curves facing the monopolistically competitive dealers.[4] ZS is the marginal revenue curve derived from ZR. YM is the marginal net revenue curve after subtracting the dealers' marginal costs other than those for raw fish, for example, marginal expenditures for boxes, labor, and so on. That curve falls just slightly faster than ZS because the latter expenditures may be assumed to rise slightly as output goes up. The dealers' market demand curve for raw fish is WN, derived from YM. Every point on WN equals two-fifths YM because we are assuming

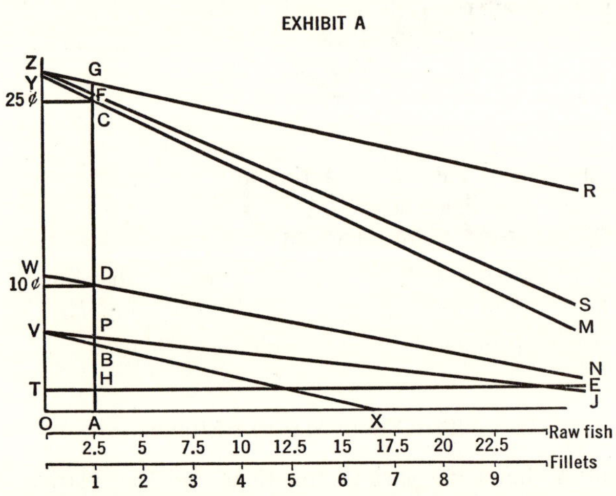

EXHIBIT A

a 40 per cent yield.[5] Hence OA (2.5 units of raw fish) times AD (10¢) equals OA (one unit of fillets) times AC (25¢). Put another way, OA represents fillet units when used with YM, and OA represents raw fish units when used with WN. The latter curve is the

---

[4] The dealers are merchandising packaged fish in the main. Hence dealer output is trade-marked or differentiated so that they are really "monopolistic" competitors. See in this connection E. H. Chamberlin, *The Theory of Monopolistic Competition* (Harvard University Press: Cambridge, Massachusetts, 1948).

[5] See page 179, this Supplement. We might point out that the difference between ZS and YM represents the marginal cost per pound of fillets. It might be converted into marginal cost per pound of fish by dividing it by 2.5 times the relevant quantity of fillets.

dealers' market demand curve facing the fishermen and vessel owners for raw fish.[6]

VJ is the demand curve facing the fishermen. They receive 60 per
cent of the catch value after deduction of certain joint expenses of
the voyage. If we disregard these joint expenses, then we may derive
VJ by making every point on that curve equal to three-fifths WN.[7]
VX is the fishermen's marginal net revenue curve. TE is the fishermen's marginal cost curve.[8] They pay the operating expenses of each
fishing trip (oil, food, etc.) and pay for exactly what they use.[9] Two
opposing forces are at work to influence the shape of the marginal
cost function. First, if one were to give some weight to the fishermen's
relative preference for income and leisure, then it would be necessary
to consider the function a rising one, for increased catch size depends
to some extent on longer trips. Second, if one were to keep in mind
that the fishermen must spend upwards of two days in sailing to the
grounds, then the marginal cost curve might be portrayed as declining.[10] In showing the curve as horizontal in Exhibit A we have assumed that on balance these factors more or less cancel out.

To sum up, let us suppose that on a given morning the demand
conditions at the port are as set forth in Exhibit A and the fishing
fleet lands a total of OA pounds of raw fish.[11] AH would be the fishermen's marginal cost; AB would be their marginal revenue; AP would
be the portion of the raw fish price they would receive; AD would be
the price bid for raw fish. In terms of fillet units, AC would be the
dealers' marginal net revenue; FC would be the dealers' marginal
cost; AG would be the wholesaler fillet price offered the dealers.

Clearly under the demand conditions posited, the fishermen would

[6] It should be noted that the yield factor might have been taken into account
on the quantity axis instead of the price axis.

[7] The joint expenses generally are small in relation to catch value.

[8] This is the "aggregate" marginal cost curve for that part of the fleet landing
on a particular day.

[9] In connection with food bills, precision might require that only the portion
of food bills be counted which exceeded what it would cost the fishermen to eat
at home. If we take into account the arduous duties of the men, however (and
the fact that "away from home workers" usually get expenses), then it does
not seem out of order to count the whole food bill.

[10] The fishermen spend at least three days per round trip as travel time.

[11] For simplicity we are assuming here that there is involved only one species
of fish of a given quality. Later we will drop this assumption.

not maximize their wage bill. Their marginal revenue would exceed their marginal cost at output OA. This, of course, should stimulate the fishermen to increase their catch on subsequent trips should the same demand conditions continue to prevail.

It is important to bear in mind that the dealers, because of the phenomenon of filleting, must constantly think in terms of both fillet units and raw fish units. This may be seen from examining the following table:

Raw Fish Price (per lb.)	Yield	Fillet Price (per lb.)	Pounds of Raw Fish	Pounds of Fillets
9¢	40%	22½¢	1000	400
10¢	40%	25¢	1000	400
11¢	40%	27½¢	1000	400

As the table reveals, changes in raw fish prices per pound make a much more considerable difference in the rate per pound the dealer must collect for his wares from the wholesaler. As a practical matter, then, the dealer can't say, "Oh, well, I paid a cent a pound too much for fish this morning but I can 'get out of the hole' by picking up a cent a pound more for fillets somehow." On the contrary, he will have to pick up 2½¢ more per pound for the fillets. If we keep in mind that the demand for fillets (and hence the demand for raw fish) is generally quite elastic, then it is not too difficult to understand that some or all of the dealers who may have mistakenly "bid too much" have a strong incentive to make the amount they actually pay when they finally settle up with the fishermen something less than the original bid.

The situations set up in the following sections are models of the fishermen-dealer relationship. They seek to explain how, under stated assumptions, diversity of interest between the fishermen and dealers might produce bitter clashes, restrictive union policies, or both. It would be well to remember, however, that any given clash between the parties, or any given set of union policies, may result merely from the fact that income under the lay just does not meet the fishermen's idea of what they think it ought to be in view of the work they do. The fishermen do not think in terms of demand and supply curves. They do have rather definite ideas of the income their efforts ought to bring them.

WHY THE PARTIES CLASH IN THE PRIMARY MARKET

CASE I: THE PARTIES DIFFER ABOUT THE
LOCATION OF THE DEMAND CURVE

Exhibit B portrays the situation in the primary market. VS and PW represent two different demand levels. VX and PY are the respective marginal net revenue curves for the fishermen. TE is the fishermen's marginal cost curve.

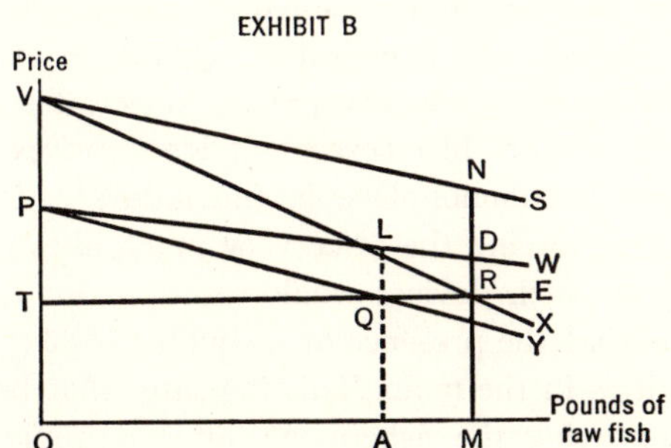

First, let us assume that with normal fishing effort using equipment in force, and with no limits in effect, the fishermen land quantity OM. Let us assume further that OM constitutes more or less "normal" landings. The plant capacity in the port is geared to handle most efficiently about that quantity. Suppose, too, that demand is PW when the landings OM materialize.

1. The fishermen clearly will fail to maximize their wage bill if the dealers pay price DM. The fishermen's marginal revenue curve will cut DM well below the point at which marginal cost and revenue would be equal. Assume further that in landing OM the fishermen had expected the demand to be VS, or approximately that, and therefore were looking forward to a wage bill of OM multiplied by RN. In effect, the fishermen will have been expecting a fish price of NM, not the price DM. Within the framework of Exhibit B, the dealers will stand firm at price DM — a price which would not even give the fishermen a wage bill equal to LQ times OA; they could have had the latter simply by landing less fish. No wonder such a situation has the makings of a good fight. In essence, this is what happened at Boston

Fish Pier in February, 1947, when the Massachusetts Attorney-General intervened, through the courts, to restrain the fishermen from insisting upon higher prices than the buyers would pay.[12] Barring such an intervention, the final price would depend upon the willingness of the parties to fight. But in view of the fact that the fish are apt to spoil if the fishermen hold out too long and in the light of demand realities as we have assumed them, it would perhaps be closest to the truth if we were to assume that the final price would be about DM.

At least two things should be noted here. First, such a case helps explain why alleged "shortweighting," claims of excessive ice and trash in the haul on the dealers' part, and periodic refusals by the latter to take any more fish (except at lower prices than originally bid), have been a commonplace in the industry. If the fishermen should succeed in forcing the price very much above DM, the pressures to resort to such devices would multiply for the dealers. The second thing is that the presence of a slightly higher demand washes out the difficulties in the main. This indicates that better demand is the key to improving the fishermen's lot and minimizing frictions. Precisely such a more favorable demand generally featured the Lenten season, at least until quite recently.[13]

To return to Exhibit B, let us suppose that when the fishermen depart for another set of trips they correctly estimate that demand PW will prevail when they return. Under such circumstances, union policy which holds output to OA will maximize the fishermen's wage bill.

2. Yet the story is still not complete even if we assume the fishermen are successful in implementing such a limit policy. (We have prescinded from the rather awful difficulty in practice which multiplicity of species presents.) The dealers will scream because of the limits. Amount OA may maximize the fishermen's position under existing demand conditions, but amount OM will be the quantity the

---

[12] See Chapter V.

[13] Until recently the movement of prices and cold storage holdings tended to indicate the more favorable demand conditions prevailing during Lent. A recent study, however, failed to find through questioning householders that this is still as true as in the past. See W. H. Stolting and M. J. Garfield, "Fish and Shellfish Preferences of Household Consumers — 1951," Fishery Leaflets 407 and 408, U.S. Fish and Wildlife Service, Washington, D.C., May and September, 1952.

dealers are geared to handle most efficiently (since we have assumed that landings of OM are about "normal"). Indeed, dealer protests in Boston in the past have amplified because in that port the larger dealers have large investments in vessels to insure that the catch will be large.[14]

The successful application of limit policy here, then, is from the fishermen's point of view but making the most of a none too desirable situation. Much better incomes would be available for all if demand could only be shifted to the right.

CASE II: THE PARTIES DIFFER ABOUT THE STEEPNESS
OF THE DEMAND CURVE

Exhibit C sets up a basic conflict situation where there are two demand curves of differing steepness. Assume again that the fishermen bring in a "normal" catch of OM (no limits in force). They expect

EXHIBIT C

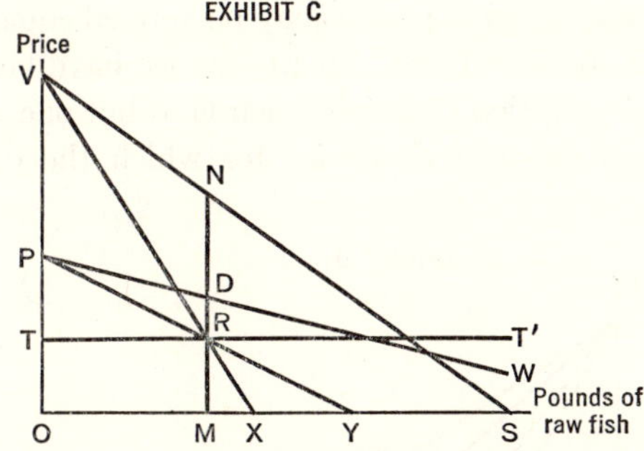

demand to be VS. In reality it is PW. Consequently, the fishermen receive a wage bill of OM times RD instead of the anticipated OM times RN. The same skirmish over prices may follow as we have described in the previous section. But here, after the hostilities cease and the fishermen look to the next auction, they find themselves powerless to change the situation through the use of "limits." (That

---

[14] We assume here, of course, that at this time the dealers have determined capacity with an expectation of OM landings or better. If we were to pursue this point further, we should have to take into account the fact that under such circumstances the demand curve might slope downward more steeply beyond OM.

is, providing they diagnose the source of their difficulty correctly.) Bringing in any less than OM will hurt, not help, the situation. In this case the fishermen's only recourse is to take steps to raise the demand curve. This case is, of course, in part the same as that set forth in Case I because VS is in effect further to the right through most of its length than PW. But under the circumstances of Case II it seems apparent that if the fishermen think the relevant curve is not only higher but also steeper, the frictions may intensify.

Before turning to the next case, let us note that if demand facing the fishermen tended to vary in somewhat random fashion from trip to trip, then the use of limits on output would seem clearly inappropriate. The fishermen have no crystal ball. They would run the serious risk of having output limitations operative at precisely the times when the limits would result in landings too low to maximize the wage bill.

CASE III: WHY THE FISHERMEN MAY LIMIT CERTAIN SPECIES

The analysis employed in Case I may be applied equally to the case of an individual species of fish. Up to now we have been proceeding on the general assumption that fishermen land but one species of fish. In reality, they land several species for which the demand curves may vary.

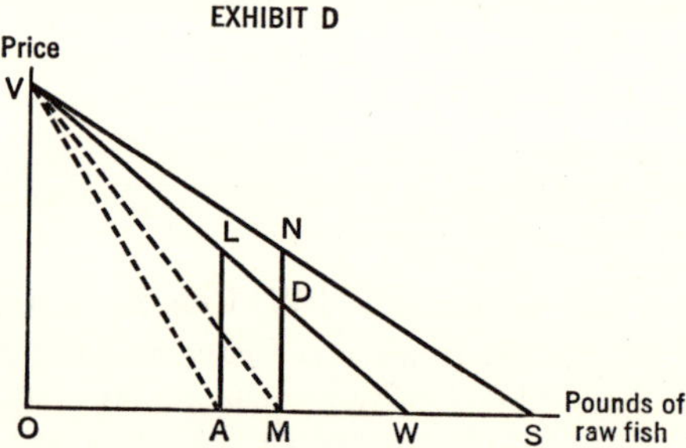

There are species, for example, for which the demand is relatively steep but limited. An example might be pollock, which seems to be eaten by only those relatively few people who know its real worth and taste. Exhibit D affords a vehicle for analyzing what takes place here. Let us assume that because the catch of these species is a sort of joint product of dragging for more popular varieties, the fishermen consider no costs attached to quantities secured.[15] In terms of Exhibit D, the fishermen would attempt to land OM if the demand is VS, and OA if the demand is VW. The relevant marginal revenue curves are VM and VA respectively. Of all the cases in which "limits" might be employed, the case of fish facing a steep, but limited demand would seem the most defensible. As the union's leader puts it, "I don't see why the men should work for nothing," bringing in additional fish that adds nothing to their incomes — and we might add — that actually cuts them.

Yet, even here, there is room for expansionist thinking. We have pointed out in an earlier chapter policies which might be pursued to promote sales of such lesser known varieties.

CASE IV: THE SPECIAL CASE OF BOSTON AND THE SELLOVER

Situation I: The Dealers Do Not Have Demand Curves Which Distinguish Quality

Chapter V has pointed out the bitterness which has featured the operation of the sellover in the Boston market and has described the essential features of that device.[16]

Analytically, the sellover may be presumed to function in two kinds of markets: first, a market in which the dealers really have only a "one quality" demand curve (as the fishermen have charged vociferously) and, second, a market in which the dealers have two distinct demand curves, one for "No. 1" and one for "No. 2" fish. In this section we shall examine the former case. Later we shall look at the "double demand curve" market.

Exhibit E sets up the relevant functions for a given day at the New England Fish Exchange. Suppose that when the market opens, VS is the demand presented by the dealers to the fishermen and VX is the fishermen's marginal revenue curve. Assume, too, that VS is the dealers' only demand curve; they do not have a second relating to a different quality of fish. Let OA be the catch with no limits in force.

The dealers will bid AB. The fishermen's marginal revenue will equal their marginal cost. All will be "sweetness and light." It usually

---

[15] In such a situation the fishermen might charge all costs against the more popular species. Such a policy would increase landings of lesser known fish.

[16] See section on the "sellover."

is in the first sales market, and this tends to indicate that on opening bids the demand curve is high enough and elastic enough so that the fishermen's marginal revenue equals or exceeds their marginal costs for the landings of that day.[17]

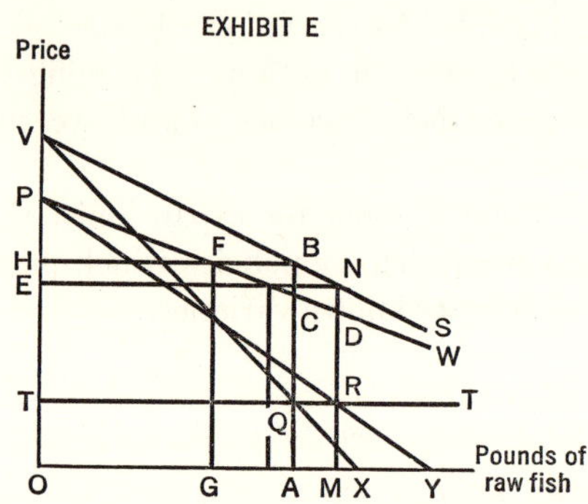

Under such assumptions what forces might lead to a sellover? Is not a sellover impossible if we assume, as we have in this situation, that the dealers do not have two distinct "quality" demand curves?

Even under our assumptions two things may still bring the sellover into play. First, with no change in the true market demand curve facing the dealers, the demand curve presented by the dealers to the fishermen may still change after the first sales market has been completed. The latter may have been a "trial demand curve" — a sort of sounding board — through which at least some of the dealers purchasing in the first sales market may have discovered that they misjudged the true demand facing the port for fillets and consequently bid too much. In terms of Exhibit E, the "true" demand curve which the dealers should have presented to the fishermen might be PW, not VS. This would make quite a difference in the worth of fish to the dealers. If the dealers were not very careful in bidding in the first sales market and relied on the availability of the sellover, there might well be a sharp difference between the price the dealers actually bid and the price they should have bid in the first sales market.

Second, even if the dealers judged demand correctly and bid on the basis of that demand in the first sales market, there could still be an

---

[17] See, however, the exception noted under Case I above.

unfavorable change in demand based upon some unforeseen change for the bad in the demand for fillets, after the first sales market had been completed but before all the fish were removed from the vessels.[18]

To illustrate how the above factors might involve the sellover, assume that the new demand curve resulting from their operation is PW. Faced with this situation, the dealers would insist upon a sellover when OG quantity had been removed from the vessels. The reason is simply that under the new demand conditions the dealers could not afford to take any more fish than OG at price FG (equal to AB). The quality factor becomes involved because to claim lower quality would be the only way to obtain a sellover. The dealers must claim that the fish left, after OG are taken out, are No. 2 and might so argue regardless of the objective quality of those fish. The fishermen, convinced for years that there is only one demand curve and that the dealers do not distinguish quality when they cut fillets for the trade, will point to the original bid and insist that all the fish move at that price. The fishermen have always charged that the only reason sellovers take place is that the buyers either have made mistakes in bidding in the first sales market or demand has changed for the worse. In either case, the fishermen and their union representatives feel strongly that the dealers ought to take the consequences as "tough luck." The dealers, of course, steadfastly insist this is not so; they attribute all sellovers strictly to quality considerations.

Let us suppose that the dealers are successful in their stand and a sellover takes place. (That is, the fish in question are sufficiently poor in quality to satisfy the Exchange Manager or his appointee that a sellover is justified.) What price will the fishermen receive for the balance of the catch GA? In the absence of collusion among the buyers, the sellover price will be about AC. With collusion, of course, the price could be anything from A to C depending upon the cohesion of the combine. For at least 15 years the union has insisted there is such collusion but has never been able satisfactorily to prove the charge.[19] Under competitive conditions the price should approach AC

---

[18] These forces could move in such a way that one offset the other and no sellover followed. For example, if the dealers overestimated true demand in the first sales market but demand changed in a favorable direction before the fish were taken out of the vessel, there would be no sellover.

[19] The Massachusetts State Federation of Labor (AFL) had before it in 1938 a resolution asking federal intervention on this and other matters.

because that is about the marginal worth of the fish to the dealers; [20] that is the price they would have bid in the first place in the first sales market if they had operated on the correct demand curve. The case is one of monopolistic price discrimination (price shading under the demand curve) with the fishermen the unwitting monopolists. Hence, under the conditions we have assumed, a paradox results. The fishermen claim they have been badly "gypped" because the whole load does not move at price AB. Yet, under the real demand conditions prevailing, they may have received more because of price discrimination. The fishermen, however, would never be able to see it quite this way, and perhaps rightly so. For any change in demand, after the first sales market ends, sets up the same kinds of situations we have considered as Cases I and II — differences between the parties over the location and steepness of the demand curve. For the fishermen the problem is how to get AB for the whole haul, and anything less they undoubtedly will continue to consider "burglary" by the dealers so long as the present form of the sellover continues to exist.

There is another feature of the sellover arrangement which merits explanation here. Suppose that when a given dealer asks for a sellover that the dealer vessel-owner from whose boat fish are being removed sides with the buyer rather than the fishermen. Why should the vessel-owner do this? Is he not, as the vessel-owner, as interested in maximizing fish prices as are the members of the crew? He may not be; he has to split 60–40 with the fishermen every cent received as vessel-owner. Every cent he saves as a fish buyer, however, is his own. Therefore, in a given situation, where dealer vessel-owner A as a buyer is seeking a sellover in withdrawing fish from a boat owned by a dealer vessel-owner B, owner B may side with buyer A. This would not mean that they were openly in collusion. It would only constitute implicit recognition by the seller (dealer-owner B) that on that or some other day the tables might turn. He would be the buyer and not want the opposition of the dealer-owner from whose boat he was withdrawing fish. Since most of the vessels operating have been owned by the more important dealers, this kind of "round robin," without anyone doing any more than following his own best interest,

---

[20] The dealers would probably be willing to pay something less than AC because, having already bought a large quantity of fish, the marginal worth of the remainder might now be less to them.

may very probably have been as common as the union claims. The important thing is that it does not have to be "collusion" as the union has insisted, but is rather a pattern of action flowing almost naturally from the present market setup.[21]

Lest we leave the impression that all the dealers all the time have uniformly supported the sellover in its present form, we hasten to add that a minority has opposed it in principle because they think it does the port more harm than good. They point particularly to the bad feeling it generates. Then, too, one dealer especially has insisted that more than once he has bought fish at higher prices in the first sales market, only to be undersold in the fillet trade by some other dealer who obtained his fish more cheaply after a sellover.

The fishermen may not be nearly so badly off under the sellover arrangement in Boston, however, as they have been in Gloucester and New Bedford. In those ports there is no sellover, but unfavorable shifts in demand and mistakes in bidding do occur. As a consequence, without the sellover the dealers turn to alleged claims of excessive ice and trash or alleged short weighting to get the same results. Clearly the Boston system is preferable.

Tarrying a further moment on Exhibit E, we should note that if the demand should become PW instead of VS, the fishermen would not be well advised to institute limits. The new situation would be somewhat like that in Exhibit C, for the new demand curve is flatter than the old. Actually, demand curve PW would require that the fishermen increase landings to maximize their wage bill. Unfortunately, expanded landings from OA to OM would leave the fishermen worse off than with landings OA and demand curve VS, but this just demonstrates once again the terrific stake the fishermen have in a high level of demand. This makes it abundantly clear that the Fishermen's Union under the present wage system cannot neglect the problem of raising and maintaining demand. It seems clear, also, that the fishermen need a union, for so great are the pressures on the dealers to secure fish at the lowest possible price that the individual fisherman or crew is no match for the dealer.

Now, let us look at the sellover from another angle.

---

[21] The dealer vessel owners are the largest buyers of fish in Boston, with the exception of the chain stores, which do not own boats.

Situation II: The Dealers Have Demand Curves Which Distinguish Quality

In the preceding section, we went ahead entirely on the assumption that market forces, not quality standards, governed the sellover. We regarded quality claims as incidental to the working out of market factors. We did this to show clearly the influence of such factors in sellover operations.

In reality, the dealers undoubtedly have quality preferences. Hence they have a different demand curve for "No. 1" and "No. 2" fish respectively. One way to approach this situation is to assume what is not true at present and construct a model in which (a) quality determination is objectively handled and (b) there are two reasonably distinct markets, one for each grade of fish. The model is set forth in Exhibit F.

EXHIBIT F

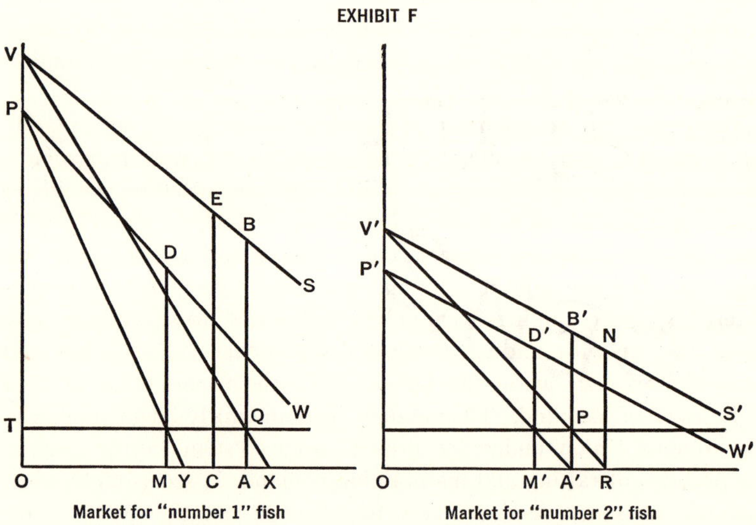

Assume that on a given day landings consist of OA, represented by the sellers as No. 1 fish, and OA', represented by the sellers as No. 2 fish. In the first quality market if VS is the demand curve, OA would sell at price AB. The fishermen's marginal cost and marginal revenue curves would intersect at Q and the wage bill for this type of fish would be maximized. Similarly in the No. 2 market, landings of OA′ would sell for A′B′ and the wage bill would be maximized.

Let us focus attention on conditions in the first sales market. In that market, the quantity of fish represented to be first quality would be the fishermen's estimate and would depend upon their judgment. If the buyers agree with the estimate, no difficulties will ensue. Suppose, however, that after one or more of the buyers have withdrawn a certain quantity of fish — say OC in Exhibit F — that the objection is made that the fish are no longer "No. 1." An impartial party, with or without fixed standards, would have to decide the issue. If he rules the fish are still "No. 1," they will continue to come out at the "No. 1" price. But if he agrees they are "No. 2," they will be sold over in the "No. 2" market. Assuming that CA in the "No. 1" market equals A′R in the "No. 2" market, then with competitive bidding the price for A′R should be about RN. This is a case of price discrimination of the kind we described in the preceding section.

The other curves in Exhibit F show a shifted market situation. The changes are similar to those posited in Exhibit E. Both new demand curves (PW in the "No. 1" market and P′W′ in the "No. 2" market) are lower than the curves assumed at the outset. Clearly, both of these shifts in demand, if unanticipated, would mean substantial embarrassment to the dealers. Hence the dealers would be under pressure to make their opening bids eminently realistic. With the type of system we have been describing here (which does not exist at present), quality as a force in determining prices would be carefully delineated; the bulk of the fish would be sold on a graded basis. In the opening auction there would be frank acknowledgment that the total catch contained certain quantities of "No. 1" and "No. 2" fish. If the market changed for the worse after the opening auctions were completed, that fact alone could not possibly by itself bring the sellover into operation. Quality conditions would have to warrant a sellover.[22]

Indeed, the clearcut "double market" arrangement set forth in Exhibit F departs sharply from present practice on Boston Fish Pier.

[22] Put another way, existing arrangements in Boston make it possible for both market forces and/or quality considerations to stimulate demands for a sellover. Under the proposed arrangement, quality differences would be openly taken into account. There would, then, be nothing to prevent equal prices in both markets ("No. 1 and No. 2"), if total fish demand were strong enough.

Under the present system, the dealers bid in the first sales market for all the fish available, much as if in fact they had only one demand curve, say VS.[23] If a sellover is later demanded, no one really knows whether market forces, quality forces, or both are at work. Quality reasons will be given, of course, to legitimatize the plea. Under such circumstances, the parties do indeed have a "built-in war" in the making. Concomitant price disappointments can and have moved the union to adopt restrictive policies which, on balance, may well have done both the fishermen and the port little good.

The clearer arrangements of the kind we have predicated in Exhibit F might not induce the union to cease employing "limits," for the fishermen still might try to "balance supplies" in the first and second grade markets so that the wage bill in each would be maximized. What the union would do on these matters would, of course, depend upon future union attitudes as to the best way to solve the members' income problems and the union's security problem. But one thing does seem reasonably certain. A clearcut "double market" arrangement would eliminate a major source of friction which has helped keep the parties in a state of armed truce for over fifteen years.

[23] Perhaps the only clear-cut exception to this at present is the case presented by Georges haddock. Haddock from the latter bank, supposedly the finest obtainable, are identified for sale separately from haddock caught on other banks.

INDEX

INDEX